AF805803

GLASS BODIES

Cultural Memory in the Present

Hent de Vries, Editor

GLASS BODIES

Fantasies of the Human in Cultural and Literary History

Elena Fabietti

STANFORD UNIVERSITY PRESS
Stanford, California

Stanford University Press
Stanford, California

© 2027 by Elena Fabietti. All rights reserved.

No part of this book may be reproduced or transmitted in any form or by any means, electronic or mechanical, including photocopying and recording, or in any information storage or retrieval system, without the prior written permission of Stanford University Press.

Library of Congress Cataloging-in-Publication Data
Names: Fabietti, Elena author
Title: Glass bodies : fantasies of the human in cultural and literary history / Elena Fabietti.
Other titles: Cultural memory in the present
Description: Stanford, California : Stanford University Press, 2026. | Series: Cultural memory in the present | Includes bibliographical references and index.
Identifiers: LCCN 2026024924 (print) | LCCN 2026024925 (ebook) | ISBN 9781503647473 cloth | ISBN 9781503648319 epub
Subjects: LCSH: Human body in literature | Homunculus (Fictitious character) | European literature—Themes, motives | Human body (Philosophy)—Europe—History | Human anatomy—Models—Europe—History | Glass manufacture—Europe—History
Classification: LCC PN56.B62 F33 2026 (print) | LCC PN56.B62 (ebook)
LC record available at https://lccn.loc.gov/2026024924
LC ebook record available at https://lccn.loc.gov/2026024925

Cover design: Abby Weintraub
Cover sculpture: Andrea Tagliapietra, *Kneeling Woman (Donna inginocchiata)*, ca. 1955, solid glass, Murano. Reproduced by permission of the artist. Photograph by Clevedon Sale Rooms.

The authorized representative in the EU for product safety and compliance is: Mare Nostrum Group B.V. | Doelen 72 | 4831 GR Breda | The Netherlands | Email address: gpsr@mare-nostrum.co.uk | KVK chamber of commerce number: 96249943

Printed and bound by CPI Group (UK) Ltd, Croydon, CR0 4YY

To Jonas and Lukas

Contents

Acknowledgments

This book is the fruit of many years of work and engagement with questions that, over time, have shifted, changed, clarified, at times even led me elsewhere as I kept thinking and talking about them in conversations rich with ever-new insights. Across these years, I have shared thoughts, words, and images with interlocutors near and far, whose contribution to this book I wish to acknowledge here.

The first idea of this book took shape during my time at the Humanities Center of Johns Hopkins University, a unique site of intellectual inquiry, discovery, and discussion. I am especially grateful to Leonardo Lisi for years of formative mentoring that have left a deep imprint on my work and to Hent de Vries for his invaluable advice throughout. During my time in Berlin at the Zentrum für Literatur- und Kulturforschung, I developed some of the book's starting points through insightful conversations with Sigrid Weigel. I received precious feedback on earlier stages of my work from Jacques Neefs, Drew Daniel, and Evelyne Ender. Bernadette Wegenstein generously supported my research with a fellowship from the Johns Hopkins Center for Advanced Media Studies. In the past years working in the Department of German Literature at the University of Regensburg, I was able to continue this project thanks to the support of Marcus Hahn. I am especially grateful to my editors at Stanford University Press, Caroline McKusick and Erica Wetter, for working on this project with trust and care and to the anonymous peer-reviewers, who provided me with constructive and helpful comments. I also owe special thanks to Elizabeth Brogden for her help in preparing the manuscript and to Joe Abbott for his careful editing.

The years I have spent moving and switching languages through the

United States, Germany, and Italy, through seminars and conversations, among mentors, colleagues, students, and friends, now appear to me as an almost boundless cosmos of exchange—and they have left countless traces, visible and invisible, far too many in fact to fully name. I wish at least to thank here, whether for their insights, support, criticism, or uplifting conversations, in alphabetical order: Rainer Barbey, Katharina Boehm, Katarzyna Bojarska, Katie Boyce-Jacino, Claudia Brodsky, Anne Brüske, Martijn Buijs, Gülru Çakmak, Alice Christensen, Jeffrey Di Leo, Tarek Dika, Simon During, Michael Fried, Matteo Galli, Stefanos Geroulanos, Zoë Ghyselinck, Joseph Griesbeck, Bart Haensel, Alena Heinritz, Jonas Hock, Irina Hron, Christian Jany, Katarzyna Jastal, Stefan Laube, Ruth Leys, Michael Lorper, Simone Marchesi, Elisabetta Marino, Paola Marrati, Omid Mehrgan, Anne Eakin Moss, Yi-Ping Ong, Orna Ophir, Massimo Petrozzi, Krzysztof Pijarski, Antonio Prete, Denise Reimann, Christian Reiß, Aviva Rot, Avraham Rot, Julia Schatz, Daniel Schwartz, Anna Seidel, Benjamin Stein, Antonio Trampus, Barbara Visser, Margarete Vöhringer, and Stefan Willer.

My family in Italy has cheered me on and has always remained a fixture of my life. My late father, Ugo Fabietti, has been an invisible guide throughout the years of completing this book.

Frederic Ponten has supported me from beginning to end, offering thoughtful feedback, patient assistance, and most of all unflagging interest and humor in our ongoing exchange.

My children have been, and are, my greatest source of inspiration, and not a few of the illuminating conversations I have had with them found their way into this book, which is dedicated to them.

GLASS BODIES

Introduction

Thus it will be: A figurine, transparent,
Stands on an earthen dish that's clean and wide,
And like a snow-white winter squirrel pelt
A girl leans over wax and looks inside.

MANDELSTAM, *Tristia*[1]

We also say of some people that they are transparent to us. It is, however, important as regards this observation that one human being can be a complete enigma to another. We learn this when we come into a foreign country with entirely foreign traditions; and, what is more, even given a mastery of the country's language. We do not *understand* the people. (And not because of not knowing what they are saying to themselves.) We cannot find our feet with them.

WITTGENSTEIN, *Philosophical Investigations*[2]

A transparent person is a common figure of speech, as Wittgenstein reminds us. A transparent person is somebody we can perfectly understand—transparency the comforting and reassuring opposite of enigma.

The present research has sprung from a similar investment in common sense: why is it that transparent persons, hearts, and minds, crowd the linguistic imagination of our speech about ourselves and others? And further along these lines, why is it that transparency appears to give us comfort and to be such an appealing or productive property that we tend to apply it to a wide range of subjects and situations, speaking of transparent persons, behaviors, institutions? Have we always been attracted to the imagination of transparent people or, rather, haunted by it? Are we drawn to this fantasy as the girl in Mandelstam's poem is to the divinatory figurine she wants to look through?

The fable of Momus, first told by Aesop in the sixth century BCE and retold some seven centuries later by Lucian, would seem to indicate as much:

> The story goes that Zeus, Poseidon and Athena were arguing about who could make something truly good. Zeus made the most excellent of all animals, man, while Athena made a house for people to live in, and, when it was his turn, Poseidon made a bull. Momus was selected to judge the competition, for he was still living among the gods at that time. Given that Momus was inclined to dislike them all, he immediately started to criticize the bull for not having eyes under his horns to let him take aim when he gored something; he criticized man for not having been given a window into his heart so that his neighbor could see what he was planning; and he criticized the house because it had not been made with iron wheels at its base, which would have made it possible for the owners of the house to move it from place to place when they went travelling.[3]

Momus's reproaches are informed by pragmatic wisdom and criticism; to expect houses to be mobile and bulls to look straight ahead is a stance no less practical than to expect humans to be transparently comprehensible and not to pose a constant enigma to each other. In the Greek text, however, a window proper is not mentioned: the phrasing indicates doors (θυρωτὰ) and an opening on the chest (ἀνοικτὰ τὰ στῆθη) or, in a different version, just the exterior location of thoughts (ἔξωθεν), so that they do not remain hidden (λανθάνωσιν);[4] in other words, no window is required for the thoughts to be exposed on the outside, so that one could easily think of other imaginative devices allowing for such exposure.

Some seven centuries later, Lucian retold the tale in a slightly modified fashion in his philosophical dialogue *Hermotimus, or Concerning the Sects*:

> The story goes that Athena, Poseidon, and Hephaestus were quarrelling over which of them was the best artist. Poseidon modelled a bull, Athena designed a house, while Hephaestus, it seems, put together a man. When they came to Momus, whom they had appointed judge, he examined the work of each. What faults he found in the other two we need not say, but his criticism of the man and his reproof of the craftsman, Hephaestus, was this: he had not made windows in his chest which could be opened to let everyone see his desires and thoughts and if he were lying or telling the truth. Momus, of course, being shortsighted, held such notions about men, but you have better sight than Lynceus and, it seems, see through the chest to what is inside, and everything is revealed to you, and you know not only what each man wants and thinks, but also who is better or worse.[5]

Besides introducing, as maker of humans, the craftsman-god Hephaestus, Lucian adds to Momus's reproach the lack of windows (θυρίδας) on the human chest that would hypothetically make it possible to recognize the desires and thoughts of human beings (ἃ βούλεται καὶ ἐπινοεῖ) and, most of all, whether somebody is lying or telling the truth (εἰ ψεύδεται ἢ ἀληθεύει). In the context of the dialogue, Skeptic Lycinus-Lucian is criticizing Stoic Hermotimus for the superficiality and dogmatism of his philosophical choices, based on questionable preferences dictated by fashion. Making fun of his naively Stoic interlocutor, Lycinus ironically says that Momus must have been shortsighted (ἀμβλυώττων) to take the problem of interiority so seriously and to place such a great importance on a window on the chest. On the contrary, what Lycinus is implying is that Momus always saw deep enough into the predicaments of human beings and their almost impossible knowledge of themselves, bound as they are to lack of transparency, their interiority buried under layers of flesh and words alike.

From Lucian on, a window on the chest has come to represent a radical fantasy of an alternative way to imagine human beings, overcoming the actual conditions of their embodiment and the corresponding cognitive limitations. Every subsequent fantasy of a transparent human being refers in a more or less direct way to this mythological representation as to a sort of archetypical fantasy. More important, Lucian's text has introduced, in the idea of exposing human interiority and making it legible, the element of mediation: a window, a pane of glass, something to look *through*.

Throughout history, transparent humans punctuate European cultural imagination: consistently present but each time meaning something slightly different. In fact, transparent humans have not always coincided with the corrective architectonic fantasy of Momus nor with the comforting fantasy of a linguistic and social space where one could find oneself, as Wittgenstein put it in his *Investigations*; quite to the contrary, they sometimes happened to embody, with their peculiar, mostly clumsy and inconvenient bodies of glass or other transparent materials, troubling and inescapable elements of the human condition. What is unequivocal is that transparent humans have consistently appeared at crucial junctures of cultural history as subjects of literary, artistic, philosophical, scientific, and political imagination, and whenever they appeared, some fundamental thinking about the human was at play—some hypothesis that shared with Momus's reproach the quality of a radical fantasy, a fundamental reimagining of the conditions of being human, a provocation.

In the course of my research for this book, I have come to think of transparent humans as thought experiments, what the German vocabulary calls *Denkfiguren*: figures that help shape and articulate a thinking process—in this particular case, thinking the conditions of being human as related to the structure of interiority and bodily as well as mental boundaries. By altering the actual conditions of being human through the replacement of flesh with a transparent body, transparent humans make it possible to test the power and reach of the other defining features of being human. If flesh were not there anymore, what would human beings look like? How would they act? Would they still be human? The figure of thought thereby becomes a true thought experiment, one that pairs with many other fictional fantasies of body-hopping, afterlife, and virtual reality, to name just a few of the discursive dimensions where the limitations and possibilities of the human are constantly tested.

This kind of speculative yet vividly imaginative interrogation sustains the different transparent human figures conjured throughout modernity and analyzed in the course of this book; each of these figures performs a thought experiment under varied cultural historical conditions. In fact, this book investigates how the historically specific conditions for the emergence of each figure—both within and across different discursive forms—render the recurring fantasy of a transparent human always something quite distinct and particular to its moment.

But what do these transparent figures look like, if they are figures of thought yet not quite figures of speech? The book explores various instances of glass bodies, a heterogeneous medial and material array of constructs that emerge at the crossroads of cultural discourses—between science and literature, craftmanship and medicine, arts and politics. All the glass bodies whose stories are told here showcase qualitative and structural differences. If Momus established the figure of a human provided with a window on the chest, transparent humans can display different features and degrees of transparency. For example, their whole body can be transparent, or only part of it; transparency can be a function of glass, thus implying a particular fragility as well, or of more resistant plastic materials exposed however to fading chromatic effects. Furthermore, what the transparent body discloses, besides the "desires and thoughts" imagined by Momus, can be such different things as the anatomic structure of the inner body or the truth of someone's soul; the transparent body can be the vehicle of a delicate, precarious, and marginalized existence or that of a normatively reassured humanity. All these possibilities are conjured and explored in the case studies examined in the following chapters.

But as much as glass bodies are distinct and heterogeneously configured, they share a common semiotic trait defining their transparency. As such, the notion of transparency is capacious enough to indicate quite different properties. *Trans-parere*, or appear-through, is the Latin medieval calque of the Greek δια-φαίνω, which defines the property of an object to show through.[6] A glass window is transparent, insofar as one can see through it, as if it were not there. Transparency would then seem to define the ability of the medium of vision to almost disappear, eliding its own material presence. But in another very common use of the word, *transparent* is also applied, through a metonymical shift, to the objects that we see transparently—that is, to what we see through a transparent medium rather than the medium itself: a transparent house is a house whose transparent walls allow us to see its interior; a transparent politics is a policy system that citizens can see at work through the allegedly transparent medium of communication strategies and institutional accessibility. The transparent humans analyzed in this work function this way: they are humans whose body-case is transparent, making it possible to envision their interiority, be it anatomical or spiritual. That is, they are not, as such, transparent media of vision, as in the first definition of transparency given above, but their bodies are. In this sense, a large number of transparent humans has been excluded from the scope of this research: invisible humans, camouflaged humans, humans who entirely disappear as they become transparent media through illusionistic, magical, or other visual effects, making it possible to see what is behind or beyond them, their backgrounds, and environs.[7]

Because the transparent humans of this research are instead those whose body-cases are transparent, the focus of my analysis will be both on the body as transparent medium and on the interiority that the transparent body makes visible. Both notions—the body-case and interiority—are historically and discursively sensitive; that is, they are constantly redefined by their respective historical moment and the discursive setup of their configuration, as heterogeneous and divergent as anatomy, physiology, and theology (to name some of such discourses) may be. Even within this seemingly simple dual structure (exterior body and interiority), however, body and interiority can actually mean a range of disparate things. Interiority can mean interior anatomy, the soul, or one's thoughts: all distinct dimensions that are highly interdependent, too, as the stories told in the following pages will often have the chance to show. The question of interiority, in other words, is posed by the discursive premises generating the figure of the transparent human, be it medical diagnostics, material economy, or metaphysical imagination. Defining preemptively the kind of interiority entailed by transparent humans would run counter to the varied historical and discursive circumstances that endow the figure at each time with its specific meaning and that summon up a cultural historical narrative.

Likewise, the body as medium is constantly redefined by broad, as well as narrow, cultural conditions. Defining the body as a (transparent) medium of vision and cognition means to mobilize historically dense notions of corporeality—for example, by looking at the history of skin understood as more or less penetrable and breakable boundary and barrier—both in medical discourses and cultural practices at large.[8] The history of the body is that of a variable body, in fact a "dynamic anatomy"[9] where paradigms of corporeality shift the very boundaries and assemblages of body parts.

The dependency of the subject of this book on the ever-changing historical conditions defining both the medium of the body and notions of human interiority has oriented this study toward the work of cultural historical contextualization. Historicist in its premises, this investigation develops through the consistent use of sources that aim to provide an accurate basis to anchor each glass body in its cultural moment. Although the heterogeneity of circumstances and discursive particulars in each distinct case of the transparent human might seem to preclude a unified study of the subject as homogeneous, this diversity is both justified and warranted by an approach to cultural history that conceives of culture as an imbricated field of interconnections, in which one cultural object is never isolated and ultimately never coincides with itself, as its center and its borders shift each time that a particular cultural operation is highlighted. In other words, the ensuing discursive hybridity of the subject "transparent humans" does ratify its authenticity as an object of cultural historical analysis rather than constitute an obstacle to the research.

Transparent humans have made several isolated appearances in scholarship informed by cultural history, never however in a continuous dedicated study;[10] this is a somehow surprising fact, considering that other forms of transparency have been the subject of a number of studies examining the broad and layered network of meanings and cultural artifacts undergirding the notion. These investigations have engendered the intellectual field of research of Critical Transparency Studies and have been dedicated to the conceptual and historical expounding of the notion, understood as mainly political but also charged with the moral and more broadly philosophical set of meanings implied by its semantics.[11] Transparency has indeed come to dominate the political and social vocabulary of our age, often becoming its positive ideological norm: transparent parliaments, transparent policies, or transparent information are all indicators of a conceptual urgency that the notion of transparency has intercepted and channeled.

Historically invoked in the nineteenth century as the desirable outcome of revolutionary practices, transparency had been the guiding element of a number of modern utopias, materializing foremost in the theory and practice of architecture. From Fourier's phalansteries, through the London Crystal Palace of 1851, to the movement of the Glass architecture of the 1920s in Germany, glass constructions have come to represent the avant-garde of social utopias; in 1929, Walter Benjamin wrote that "to live in a glass house is a revolutionary virtue par excellence."[12] Transparent architecture came to fulfill the function of a myth, as architecture historian Anthony Viddler put it: "transparency of the self to nature, of the self to the other, of all selves to society."[13] But the utopian linings of the notion of transparency are even more deeply rooted in Western culture, defining "an entire socio-political tradition centered on the (often utopian) idealization of political and legal transparency,"[14] entailing theological and epistemological dimensions extending from the Augustinian *Confessions* to Marxist aesthetics.

Later accommodated into the vocabulary of reform of democratic policies, from the era of Glasnost (гласность; inexactly translated as transparency rather than with its literal meaning of "having a voice," глас meaning "voice" or "sound") to its Western counterparts, transparency became, in the last decades of the twentieth century, the subject of a critical revision: demystified as a myth or accused of evil, its virtues were drastically put into question. If, for Benjamin, the glass house was a revolutionary scene that did away with the thick, overcharged space of bourgeois interiors, criticisms of transparency concentrated on the potentially tyrannical function of exposing one's privacy to panoptic surveillance, be it from behind glass walls or through the total sharing of private data. In the twentieth century, a diversified array of intellectual positions engaged with a critique of transparency; in particular, this criticism was championed by a lineup of French authors located in the intellectual milieu of postwar France, as Stefanos Geroulanos has extensively demonstrated in his intellectual history of the critique of transparency.[15] In his study, attending to the decades right after World

War II in particular, a various yet homogeneously interconnected net of intellectual debates is shown to converge into the critique of assumptions (and ambitions) of transparency. The intellectual debates involved include existentialism's refusal of transparency (Sartre), phenomenology's establishment of non-Cartesian gnoseologies (Merleau-Ponty), and the philosophies of identity based on alterity from Fanon to Bataille, Lacan to Foucault. Transparency's critique looms large in the second half of the twentieth century, reaching toward the end of the century the works of Baudrillard and Vattimo. Yet the political currency of transparency in Western democracies of the twenty-first century seems often oblivious of these epistemological discussions; more probably, what transparency has crystallized into its current notion is only a semantic fraction of the spectrum of meanings that the concept potentially entails.

In the present book, dedicated to concrete historical instances of bodies of glass, the intellectual history of transparency, often paired with a more materialist analysis of glass architecture, is only one of the frameworks involved. The material configuration of humans with transparent bodies shares, however, with the material constellation of glass architecture more than one feature: primarily, the ethical promise of a better, purer, more honest form of existence—the promise of Momus. But transparent bodies, once engendered as fantasies, characters, or models, can also reverse that promise into a negative dream of exclusion from the social group or of loss of humanity. That is, the location of transparency within the subjective agency of the human redirects wider claims about social, political, and cultural transparency toward questions of emotional, cognitive, and relational nature. The two dimensions—the architectonic and the human—are obviously related, yet they do not coincide and may even reveal discrepancies of scale, as well as of content.

Likewise intersecting but not coinciding are several semiotic elements defining the notion of political transparency and those defining the figures of the glass bodies described herein. The specific—and to varying degrees concrete—location of transparency within the medium of the body brings the relevance of corporeality into the very definition of transparency, orienting the research both toward cultural discourses defining the body (medicine, science, health, religion, and ethics, to mention some) and toward the material conditions of the transparent embodiment (be it early modern glass, or a window on a mechanical heart, or an alchemical vial, or a transparent plastic shell: all culturally specific objects examined in the chapters that follow).

In other words, the history of transparency as a political and social concept (and its aesthetic counterpart) and the history of transparent humans do not map neatly onto one another; rather, they reveal a fruitful divergence that this study has taken up as its own challenge and specific task. While unavoidably interacting with the questions posed and investigated by the field of Critical Transparency Studies, this book carves out a materialist line of inquiry that supplies the figural and abstract rhetorics of transparency with objects, material textures, and practical circumstances (as fictional as they may be).

Material history reclaims in this book a prominent place: each glass body under investigation involves a cultural materiality, with a strong emphasis on the medial aspects of the body and its techniques. This entails, for example, reconstructing the trade of glass and the technologies of glassblowing in early modern Europe, the insurgence of mechanical models of the body and the fashion of automata in the eighteenth century, the alchemical theory of the transformation of matter, and the state-organized practices of hygiene in early twentieth-century Europe—to name some of the main material constellations at play. But it also implies addressing material dimensions imprinted by the stamp of complex symbolic assemblages such as the humoral setup of the body in early modern medicine, the medical representation of a sentient and sensitive body through animal experimentation, the Neptunist theory of the origins of life, and the sacral and spiritual iconology of the human in the sculptural artistic representation of transcendence.

Even more radically, the material implications of transparent humans investigated here are not just a function of their contextual location; rather, they reclaim for these figures—imaginative doubles of the human—a materialist immanence attesting to the human entanglement in the world of things, a durability that spans relics and debris in a history of anthropogenic objects—a position that has been voiced in recent decades by a material turn in the humanities.[16] The transparent bodies of this book, when seen from the perspective of these revived materialist insights, can highlight alternative configurations of the human substantially entangled in its material traces, thereby relocating its agency onto a site of "posthuman convergence."[17] Seen from this perspective, glass bodies can be understood as fulfilling the fantasy of an alternatively embodied human being who, while upholding cardinal points of humanism, dislodges its own center in fundamental ways. The glass body of the transparent human—through its eschewal of flesh to the advantage of a transparent, sometimes breakable, body—announces the insufficiency or even failure of humanist mainstays (say reason, soul, subjectivity, and their placeholders, according to the cultural discourse of a specific moment) as they are assumed to operate, in the humanist paradigm, within and despite the body. The posthumanist function that may be assigned to the transparent humans examined in this book hinges on the fictional questioning of humanism enacted by these figures and will therefore be invoked intermittently as a possible narrative frame for articulating their critical function. Although a viable category to expand the interpretation of these figures, their specific form of posthumanism forgoes the immateriality of what Stacy Alaimo has called "the obdurate, though postmodern, humanisms that seek transcendence or protection from the material world,"[18] which culminate in disembodied transcendence and incorporeal permanence. The fantasy of a transparent body, on the contrary, always involves some type of material (imaginary as it may be) intervention modifying the fleshy nature of the biological human, be it glass or any other historically determined support for the attainment of a transparent corporeality. In light of this theoretical revision of the materiality of the human entanglement with its environment, which contentiously unsettles the transcendent notions of the human based on cognitive, sexed, and socially powerful dualisms (mind and body, nature and culture, flesh and spirit, etc.), the glass bodies of this book restore a provocative, fantastic, or pathologically deviant representation of the body as entangled, imbricated, absorbent, and assimilated in the environmental materiality of its situatedness.

This means that transparent humans are not free-floating, crystallized ideas but material figures of an enmeshed humanity implicated in material and discursive relations with the world. Although my research shares with the time-bound methodology of the history of ideas an interest in the recurrency of a figure of thought, what is told in these pages is not, as Arthur Lovejoy wished for, the coherent plot of a "unit-idea which the historian thus isolates [as] he next seeks to trace through more than one—ultimately, indeed, through all—of the provinces of history."[19] Rather, I will pursue the muddled, disjointed, and yet continual plot of a figure that accompanies, as its satellite fantasy, the appraisal, criticism, debunking, or dissolution of the unstable notion of the human. My enterprise is therefore equally far from the traditioned aspirations of thematic criticism, with its relentless faith in the stability and reinscription of constant *topoi* in the continuous reassembling of historical constellations.

The glass bodies of this book ultimately share what I see as a recursive function: for each distinct historical constellation, transparent humans play a significant role in questioning cultural assumptions about being human. By staging an imaginative, alternative, liminal way of being human under modified conditions of embodiment, these figures consistently defy presumptions about the boundaries, stakes, and limitations of the notion of the human, while at the same time retaining a distinct *bodiliness* that does resist the temptation of dissolving the human (or its fantasy) into disembodied configurations.[20] Furthermore, if the body is always already socially inscribed, by reinscribing the conditions of human embodiment into a body devoid of flesh, transparent bodies do not simply forgo flesh but double up on their semiotic effects: bodies of glass expose the inscription of the body (usually into flesh) as a social operation, exhibiting its procedural steps, its challenges, and its shortcomings.[21] Glass bodies are being seen and recognized as bodies that have been reembodied by subtracting flesh, thereby exposing a double corporeality.

As I argued above, the transparent human works as a *Denkfigur*—a figure for thinking theoretical issues with the help of a materially informed configuration. In a more radically embodied definition of a figure, Donna Haraway has defined this epistemological notion as an "invitation to inhabit the corporeal story told in [its] lineaments." For Haraway, "figures are not representations or didactic illustrations, but rather material-semiotic nodes or knots in which diverse bodies and meanings co-shape one another."[22] It is in this vein that in the figure of the transparent human biological, metaphysical, aesthetic, and political meanings coalesce to shape a lively knot of signification. Transparent humans concretely enact these meanings; they do not just symbolize them. The aspect of performativity, in fact, defines the agency of all the heterogeneous glass bodies of this book. As fictional figures, they inhabit a space of alternative reality, one that shares some of the features of the actual world but that retains its autonomy, providing the freedom for an experiment to unfold. As fantasies, these figures carry the load of the thinking experiment with which they are invested by acting it out in the variously narrativized space of their performance. In traversing fields of knowledge that are as distinct as they are intertwined, the story told in this book doesn't emphasize simply the figural permanence of meanings but, to quite contrary ends, the epistemic power assigned to an intermittently continuous figure of thought to produce each time an occasion for radical critique. Thought experiment and *Denkfigur*, the transparent human is most and foremost a performative, fictionally plausible version of an alternative way of being (or staying) human.

The discursive context and texture sustaining each glass body's appearance affect deeply their form and semiotic configuration: whether this figure can be imagined, touched, or looked at transforms this object beyond its nominal permanence. Under different conditions of embodiment, however, these glass bodies repeatedly provide a means for rethinking the conditions of defining humanity disciplined and arbitrated by a certain culture at a specific junction. Functionally homogeneous but materially discontinuous, transparent humans enact human life by retaining its agential contours through the structural provision of a body that lacks the specific conditions of life derived from flesh—its physical and textural limitations and its biological exposure to decay and disease. By fictionally and temporarily disentangling life from the actual *bios*, the contours of the discourses that shape life become even more visible: politics, science, religion, and various forms of the metaphysical shimmer brighter in transparent bodies than in their fleshy counterparts so that these glass bodies operate as fantasies of the human as such, questioning its very existence.

Literature in this book is a privileged access point to the cultural history of transparent humans but not the exclusive one. Whenever literature engenders these figures, it provides them with the fictional space to unfold their performativity—their lived reality. Literature, however, is never taken by itself in this study but is methodologically informed by an idea of discourses as coextensive, intertwined practices shaping each other: literature and medicine, literature and philosophy, literature and science, literature and propaganda, to name some of the different exchanges explored in the course of this work. The stories recounted in the following chapters sometimes begin with, end with, traverse, or run parallel to literary texts. The fluid movement from literary to cultural analysis and back implies, on the one hand, the assumption that literature functions as a heteronomous discourse—absorbing, elaborating, and releasing themes and tensions of the broader cultural production of discourses; on the other hand, it understands literature as affording amplified space for imaginative operations ranging from utopian to critical registers.[23]

In chapter 1, I reconstruct the case of melancholic subjects who, in early modern Europe, believed their body was made of glass. Looking closely at medical sources documenting this syndrome, as well as at the protagonist of a short story by Miguel de Cervantes, "The Glass Graduate" ("El licenciado Vidriera," 1613), I show what bodies of glass, in the cultural imagination of European early modernity, came to represent: a fictional experiment of an alternative way to be human under the circumstances of mental disorder. Embodying the fragility of their imaginary body, as well as its unique material purity (as new production strategies were implemented), early modern glass humans became the site of an imaginative reorganization of the relations between the normal and the pathological, the exceptional and the strange, power and impotence. The glass delusion, as this figure has been called, did not constitute a conclusive solution to the problem of the melancholic human but rather enacted its tensions across social, medical, religious, and philosophical competing meanings. The extraordinary body of early modern glass melancholics was at the same time imaginary and real, and their existence could cross with astounding fluidity material, medical, and fictional discourses. The chapter foregrounds an investigation of the material economy of glass at the peak of glass craftsmanship on the island of Murano in the sixteenth and seventeenth centuries and the symbolic economy hinging on it. It is therefore an eloquent case in point for the cultural investigation and methodology pursued in this book, as it shows the unavoidable dependency of symbolic economies on the materiality of culture. The chapter also makes a strong case for the complex negotiations of meaning afforded by an imaginative (in different degrees) embodiment in glass. The glass body of early modern melancholics performed in the subjective reality of its carrier a fictional substitution of the conditions of human embodiment, powerfully and painfully real in its experience, allowing the melancholic subject to navigate the predicaments of social existence in a meaningful but also pragmatic way, while also serving as an interrogation of humanist tenets at the peak of humanist culture and at the brink of its potential rupture.

In chapter 2, I move to the representation of transparent humans in the age of Enlightenment, particularly in narratives by Laurence Sterne and Jean-Jacques Rousseau: *Life and Opinions of Tristram Shandy, Gentleman* (1759–67) and *The Confessions* ([1769] 1782). This literary constellation draws directly on the myth of Momus to give life and body to fictional humans endowed with a transparent heart. In the coeval narratives of Sterne and Rousseau, the transparent heart becomes a key figure to express ideas about the unattainability of knowledge about the human being, in the first case, or about the trust in its possible achievement through an exemplary but also unique act of confession, in the latter. In an age highly invested in the task of knowing, discovering, and understanding the human being by opening up new fields and formats of knowledge, from philosophical anthropology to physiology, the figure of the human with a transparent heart played a key role in negotiating the limits and reach of this very epistemic project. The metonymically transparent humans of Sterne and Rousseau are shaped by the medical and philosophical representations of their time and draw, in particular, on a rich repertoire of available (and even fashionable) mechanical models of the human being. Their function, however, turns out to be highly critical and is not simply derivative of mechanistic conceptions; in this regard, they share their epistemic agency with the other transparent humans of the book. Once more, this doesn't mean they ultimately arbitrate the epistemic conflicts they are summoned to enact but that they present fictional strategies to articulate, represent, or reimagine the cultural tensions that feed the thinking about the conditions of being human. This chapter, in particular, shows how the visual paradigm of the scientific rhetoric of the Enlightenment failed where the knowledge of the human was the knowledge of a particular individual; the transparent human thus became the chimera of thorough and complete knowledge of a knowable human sample, thereby signaling both the ambitions and the shortcomings of the epistemic project of the Enlightenment.

In chapter 3, I scrutinize an entirely different kind of transparent human: Homunculus, the little hermaphroditic human in a flask created by alchemy in Faust's laboratory in the later, second part of Goethe's *Faust* (1832). This peculiar, diminutive human dwelling in glass possesses a clairvoyant capacity, which descends from his transparent life in the glass vial. The particular embodiment of the homunculus provides him with the same essential function of transparent humans emphasized before: a fundamental rethinking of the conditions of being human through the fictional experiment of replacing the human embodiment in flesh with a transparent body. The transparency of Homunculus is, however, distinct from other transparent humans, as it is not of its body but of its embodied existence in the vial (despite the fact that homunculi are described, in classic alchemical texts, as translucent creatures). This specific, mediated transparency of the homunculus establishes the epistemic conditions for a transparent (clairvoyant and, in general, sharper) vision of the world. Homunculi, creatures generated at the intersection of nature and art, embodied yet spiritual beings, are superior to humans in their power of knowledge yet are subjected to them. Their nature is therefore markedly liminal, as I show in the course of the chapter. In Goethe's play, Homunculus yearns for a transition to a completely human body, desiring to give up his transparency, thereby openly enacting the stakes of embracing transparency or the renouncement thereof within the conception of life pursued by the Romantic agenda of *Naturphilosophie*. This Romantic-alchemic figure, through its transformative wish, indicates a path for defining the human that is alternative to that of epistemic sovereignty—the unthwarted knowledge of reality that transparency, ever since Momus, promised. The path away from transparency and toward embodiment undertaken by Homunculus takes up the burden of the body, lightened by the ravenous desire to overcome the boundaries of individual existence by embracing flesh through the material merging into the vastness of Nature. Homunculus's glass body thereby displays an inversion of the logics of the thought experiment that confirms the disruptive power of thinking through the transparent figure and possibly achieving a conceptual relocation of the human—this time within nature.

In chapter 4, a quite different glass body is the main object of investigation: an anatomical model called "gläserner Mensch" (glass human) crafted in the 1930s by the Hygiene Museum in the German city of Dresden and later circulated around the West as a didactic tool for popular medical schooling and as an entertaining object to behold. When National Socialist propaganda designers, in the course of the 1930s, exhibited this item by recontextualizing it in their exhibitions dedicated not just to hygiene but to eugenics and race politics, they transferred onto it, through a strategic use of propaganda discourse, the new disquieting projections of a human ideal conceived as a normative prototype of human evolution. This chapter takes the transparent human model from Dresden as a limit case of what transparent humans historically came to represent. In the use of this glass body in the context of museal and propagandistic political activity, conditions of embodiment are once more exploited but without the critical, doubtful, fragile questioning characterizing the other historical instantiations of transparent bodies. This time the transparent body is implemented to programmatically impose a new normativity on the human, to efface variation and deviation, to expunge complexity and singularity. While the preceding transparent humans explicitly enacted the tensions of a culture attempting, at a specific juncture, to negotiate its conflicting or insufficiently applicable notions of the human, the fourth chapter presents a structural rupture of this epistemological pattern, replacing critique and fantasy with normativity. Yet the outcomes of this operation realign with the critical function of glass bodies throughout insofar as the symbolic complexity accumulated by the figure—and its recalcitrant materiality—reasserts their function beyond the attempted normalizing goal.

In presenting a "cultural and literary history" of glass bodies, this book relies on an idea of cultural analysis unfolding historically. Despite the historicist premise at the core of my approach, it is important to underscore how this narration is not intended to be teleological. In fact, the case of glass bodies defies narratives of body history that rely on stable inscriptions of the body as carving its own dimensions (exteriority, interiority) through varying measures of closure, porosity, or malleability.[24] Even if cultural history suggests what coordinates may underpin viable notions of the body at a certain point in history, especially as far as borders, barriers, and permeance are entailed, transparent bodies unsettle those coordinates and their stability once more through their fictional makeup. My aim is not to repeat how the human body has been historically conceived and conceptualized (though I will often return to that) but to tell the story of how humans have fantasized, imagined, constructed, and reinvented their bodies through the fantasy of a glass body. This story does not commit to showing how the arc of this historical trajectory has a purposeful direction or how it is supposed to end. The uses of the figure of the transparent human continue until today, and a further study of the forms of transparent humans adopted by certain cultural constructs and figures of speech, such as "der gläserne Patient" (the glass patient) in contemporary Germany, opens up avenues to rethink this figure under the cultural conditions of the present. In my coda , I provide an overview of the poetic and rhetorical uses of this figure from the second half of the twentieth century to the present, suggesting a contingent dematerialization of this figure coextensive with the staggering development of virtual and digital identities in both information technology and social media. At the same time that transparency as a notion guides the interpretation of many communicative functions enabled or obstructed through the digital world, the figure of a transparent body inspires or threatens forms of digital and virtual identity. Through a rapid overview of transparency's swan song in modernist aesthetics and its debunking in poststructuralist thought, I attempt to understand the glass human's simultaneous dematerialization and intermittent recurrence in critical dystopian tropes, as well as in pedagogical fables.[25]

The future uses of this figure are not predictable, nor should they be. Glass bodies have played a continuous role in cultural imagination, even if their destinies have always been the particular, unexpected, and idiosyncratic product of their unique historical moment. Joining the notion of the human as its fictional counterintuitive, ominous, or liberating counterpart, the transparent human may be said to produce a permanent critique of humanism modulated on the varying meanings that the latter has taken up and will keep upholding, playing out as its apotheosis or defeat, its celebration or collapse. As opaque as any future appears in attempts at cultural divinations, as long as alternative configurations to the human of flesh and bones are sought through the possibilities of fiction and imagination, the glass human will keep reappearing. At the same time, technological and media landscapes keep transforming horizons of visibility and invisibility, while modulating different needs and desires for transparency, especially for the transparency of the human body.[26] The girl in Mandelstam's poem is still leaning on the transparent figurine, and watching inside, although we still don't know what she is looking at.

ONE

Breakable Bodies, Sharp Minds

GLASS HUMANS IN EARLY MODERN EUROPE

The first full-fledged figure of a transparent human is a product of early modern European culture. Documented in a variety of medical sources between the fifteenth and the seventeenth centuries, it is a marginal but nonetheless crucial instance within the vast repertoire of melancholic symptoms: one of the many forms that melancholy took up in its continuous yet transformative history. At the end of this chapter, I will have shown, however, how the glass delusion is not just an episodic variation within the melancholic repertoire but rather synthesizes and condenses melancholy's fundamental tensions. In other words, the glass humans of early modern Europe represented—with the concrete texture of their fragile and transparent bodies—the cultural and existential implications of melancholy itself.

The journey undertaken in this chapter is extensive and rich, for retrieving the full cultural complexity of this figure demands a comprehensive approach. The medical accounts attesting to the pathological delusion of "people of glass" are a crucial primary source, but their references to this syndrome remain scant and anecdotal. On the contrary, a literary text from 1613, Miguel de Cervantes's short story "El licenciado Vidriera" ("The Glass Graduate"), offers both a rich description and a narrativization of this phenomenon, which can illuminate the sparse information retrievable from medical texts. In this chapter, I link Cervantes's short story to the medical and philosophical sources of the time in order to redraw the cultural map that featured the figure of the glass human being. I am particularly wary of the perils of anachronistic, metaphorical readings that have both guided and misled many previous readers of Cervantes's story, and I therefore argue for a historically embedded (or historicist) metaphoric, which draws on the significations available to the actual protagonists of what we could perhaps call, today, tales of mental disorder. After sifting through the primary sources documenting the syndrome, I present it in the context of early modern melancholy, for which I emphasize distinct and apparently contradictory lines, namely its pathological dimension but also its resourceful exceptionality. I then connect this analysis, based on established interpretations of early modern melancholy, to my findings about the material culture of glass in early modern Europe. In fact, by evoking the material history of early modern glass through practices of craftsmanship and trade, I am able to show how early modern glass humans are a concrete, material inscription of the large and comprehensive culture of early modern melancholy. The properties of early modern glass applied to a body—including breakability, thinness, and transparency—become the conditions of a physical, embodied performance of melancholy. The glass delusion thus emerges as the staging (very much real for the performer) of one of many errant forms of existence that the traditional medical category of melancholy had been responsible for diagnostically and therapeutically containing and managing.

As a result, what was until today considered a little known, rather marginal, and certainly bizarre syndrome becomes in my reading a privileged place, however liminal, to understand the complexity of melancholy. The glass human of early modern Europe functions as a paradigm of the melancholic subject—that is, an exemplary case of a mentally vulnerable and socially deviant human being who nevertheless possesses a unique power unavailable to its "normal" counterpart.

Yet this paradigmatic figure, inscribed in the cultural narrative of melancholy, deeply challenges the stability of the very cultural premises on which it relies. The glass humans of early modern Europe, including Vidriera—the protagonist of Cervantes's story that we are getting to know in this chapter—are a powerful demonstration of what the fantasy of glass bodies can do in shaking the foundations of human embodiment and, with it, assumptions about the human. In the cultural imagination of early modernity, glass humans served as a fictional experiment of another way to be human, resting on entirely different conditions of embodiment that led to new possibilities of reorganizing the relation between the normal and the pathological, the exceptional and the strange, the powerless and the powerful. In fact, the glass body of these early melancholics was not just a site of fragility and breakage but also a prosthetic enhancement that enabled the unique unfolding of cognitive skills, coming to represent what may even be regarded as a posthuman embodiment in the age of humanism.

A Man of Glass

It is sometime around the year 1560 in the Spanish region surrounding Valladolid. A man believes he has become a glass vase and therefore avoids any contact with other humans for fear of shattering. He is then locked into a room filled with straw so that he can be cured; the room is then set on fire. The man cries for help, banging desperately on the door. His guardians, outside, respond to his screams by asking him why, if he really is made of glass, wouldn't he break from all that banging. To this question he replies: "Open, I am begging you, my friends and dearest relatives, because I don't think I am a glass vase but just the most miserable of all men; especially if you will let this fire put an end to my life." This little story, which might trigger in the contemporary reader unsettling representations of medical abuse, is recorded in a book written around 1569 by Spanish doctor Alonso Ponce de Santa Cruz.[1]

At the turn of the seventeenth century, when Miguel de Cervantes was writing "El licenciado Vidriera" (one of the twelve *novelas ejemplares* published in Madrid in 1613), he could have been familiar with this case.[2] He definitely was at least familiar with the symptomatology shown by the madman portrayed in it, who believed himself to be made of glass: a curious disorder of the mind, which medical texts concerned with melancholy at the time most often reported alongside a wide range of similar delusions. The syndrome appeared in several treatises from all over Europe, well beyond the Iberian peninsula—from England, through Italy and France, to the Netherlands. The circulation (and replication) of the melancholic discourse was an encompassing European phenomenon, and so were its internal narratives, including the syndrome of the glass humans.[3]

The many medical texts of the time that refer to the glass delusion mostly exhibit a descriptive drive and a taste for what is curious and strange, eschewing the etiological and semeiological expectations that govern our current understanding of medicine.[4] Jean Starobinski called them "historiettes psychiatriques."[5] It is perhaps the fundamental incommensurability between contemporary diagnostic categories and early modern ones that explains why Cervantes's story has mostly been read outside of the interpretative frame of the melancholic symptomatology, leading to a number of enticing interpretative anachronisms.[6]

In what follows, I reintegrate Cervantes's text into the context of its contemporary medical discourse in order to make use of "El licenciado Vidriera" (which certainly remains a complexly woven literary text) as a point of symbolic access to the world of early modern glass humans. I therefore engage with Cervantes's text not just as a site of exegesis but also as an exegetical prop for reading the more inscrutable medical sources, trying to activate a double hermeneutical movement from the medical sources to the novella and back.

Tomás

Tomás Rodaja is an eleven-year-old peasant boy asleep at the foot of a tree when two students riding horses encounter him for the first time.[7] Tomás wants to study and become famous, as he tells the students, and convinced by his determination, the two wealthy young gentlemen hire him as their servant, allowing him to pursue his studies in Salamanca. The boy soon shows unusual intelligence (*raro ingenio*), excelling both at his studies and at his domestic tasks, and becomes their faithful companion. After eight years, he has made a name for himself thanks to his intelligence (*buen ingenio*), outstanding ability (*notable habilidad*), strong memory (*felice memoria),* and good understanding (*buen entendimiento*). After the years spent with his two benefactors, he accompanies them on their journey back to their hometown, but he soon feels compelled to resume his university studies, the costs of which his friends generously undertake. On his return from Malaga to Salamanca, however, he meets an infantry captain who talks him into joining the army; once more, Tomás shows his *raro ingenio,* eliciting vivid interest from his new acquaintance. When the army captain invites Tomás to join him on his travels with the army, Tomás accepts the offer, traveling to Italy with the military company but without its duties. There, he visits most of the Italian landmarks in what mostly resembles a Grand Tour trip, traveling between Genoa, Florence, Rome, Naples, Venice, and Milan; finally, content with his instructive tour, he returns to Salamanca, where he finishes his studies and eventually obtains his law degree, becoming a *licenciado en Leyes.*

It is in this Spanish town that Tomás Rodaja's mates convince him to pay a visit to a worldly, resourceful lady, as the text evasively puts it, who grows fond of him but whom Tomás ignores and rejects, interested only in the pursuit of his own education. Resenting the affront, the woman has Tomás ingest a love potion hidden in a quince, which turns out to be poisonous, leaving Tomás bedridden for six months. This episode marks the text's transition to the central section, which is devoted to Tomás's peculiar madness:

> Tomás was in bed for six months, during which he dried up, and was reduced to skin and bone [se secó y se puso, como suele decirse, en los huesos], as they say, and gave every indication of having lost his senses and although they applied all possible remedies, they cured only the sickness of his body, but not that of his mind, because he regained his physical health, but was the victim of the strangest madness ever heard of. The unfortunate young man imagined that he was all made of glass [todo hecho de vidrio], and in this delusion [imaginación], whenever anyone approached him, he would shriek, begging and pleading with coherent words and arguments, for people not to come near, because they would break him, because really and truly he was not as other men—he was made of glass from head to foot [que todo era de vidrio, de pies a cabeza].[8]

Tomás begins here his new life as a man of glass, who renames himself "Vidriera," literally a windowpane. As he wanders about town and encounters people on the street, first in Salamanca and later in Valladolid, Vidriera takes up the role of a wise and pungent satirist, passing solicited and unsolicited judgments that target most professions, from doctors to shoemakers. In this new role, Vidriera is also sought after by notable and wealthy people, who summon him to court to be entertained by his madness as well as his wit. The story dwells in some detail on Vidriera's life as a glass man and on the quick-witted dialogues he entertains with his interlocutors. After an indefinite time living as a glass man, Vidriera is cured by a friar and returns to a regular life; but because of the immoderate attention he still attracts, he has to give up the practice of law and will go off to war, where he will soon die.

A Case of Melancholy

Tomás, after ingesting the philter, is unequivocally a victim of melancholy. His symptomatology is quite straightforward: he dries up and thins—a typical manifestation of the effects of a black bile imbalance. It is true, as some critics have noted, that the ingestion of the philter seems inconsistent with the usual physiological explanation of melancholy, which mostly arises and escalates, as we will soon see, from an inner imbalance of fluids in the body; but this narrative expedient doesn't challenge the status of the delusion as a manifestation of melancholy.[9] Rather, it may posit a crucial discontinuity between the medical discourse and the literary text—a discontinuity that is productive for this inquiry insofar as it carves out a "suspension" in the logic of the disease that corresponds to the function of fantasy here assigned to glass humans.

The definition of melancholy as disease can be traced back to Greek humoral theory, which Hippocrates had articulated as an equilibrium between bodily fluids that determined health and disease.[10] Melancholy was caused by an excess of black bile, *μέλαινα χολή*, conceived as a thick dark fluid exposed to thermal and physical variations inside the body.[11]

The body of knowledge assembled by Hippocrates was absorbed and reshuffled a few centuries later by Galen, who devoted several places of his work to melancholy.[12] The Galenic definition of melancholy within the system of humors provided the basis on which people understood melancholy up to early modernity, as long as medical knowledge was still fundamentally Galenic.[13] Cervantes's medical sources followed that tradition: besides the medical book by Santa Cruz mentioned earlier, he also may have had access to the *Libro de la melancolía* (*Book of Melancholy*), published in 1585 in Seville by Andrés Velásquez, a doctor from Arcos de la Frontera.[14] Mostly a work of compilation, and a didactic patchwork of transmitted knowledge, this text was the first vernacular European text on melancholy, written at the intersection of the Galenic and Scholastic traditions.[15]

This basic axis of Galenic knowledge about melancholy, while not significantly altered, was nonetheless intersected by coexisting parallel explanations, which all participated in the assembled production of the disease and drew from different traditions, such as demonology or the mythological-astrological Saturnine motif, which is at the interpretative core of the now-classic study *Saturn and Melancholy*, by Raymond Klibansky, Erwin Panofsky, and Fritz Saxl.[16]

The doctrine of the four humors, over the course of its long history, had become inevitably exposed to transformations, corrections, intermissions, and generative confusions that led to an increasing complexity of the contents indexed under the category of melancholy. For example, it was probably Rufus of Ephesus in the late first century CE who had originally introduced a distinction between natural black bile, always present in the body in varying proportions, and the pathological product of the combustion, *adustio*, of yellow bile, which resulted in a different, more dangerous kind of black bile.[17] This distinction had produced another, crucial one—especially significant in early modern accounts—between *melancholia naturalis* and *non naturalis*, which echoes contemporary notions of physiological vs. pathological states.[18]

The merging of disparate traditions, as well as imprecisions in the transmission of the Galenic corpus, added even more variety and heterogeneity to the conditions gathered under the rubric of melancholy. In fact, as much as the range of symptoms of melancholy must be considered in its entirety, its spectrum remains strikingly vast and diverse, as is to be expected for such a capacious diagnostic category that is nominally maintained while also shifting meaning throughout time.[19] The unity of melancholy, predicated on the physiological basis of the humoral imbalance or corruption of black bile, expressed itself in an array of behaviors: from solitary to depressive, from delusional to suicidal. As argued by Michel Foucault, the unity of the category of melancholy functioned primarily as a perceived unity: "a certain qualitative coherence."[20] Rather than signaling the looseness of the notion of melancholy, the great diversity and fluidity of symptoms indexed under the melancholic complex—what Drew Daniel has effectively called "the melancholy assemblage"—is rather an indication of the deep-seated, pervasive function of melancholy in early modern culture.[21]

Some stable features, however, seem to articulate a basic structure of melancholy underlying various symptomatic constellations. The two main, stable Galenic symptoms of the pathologic imbalance of black bile are fear and sadness: *metus et moestitia* [φόβος και δυσθυμία]. As such, they constitute a defining binary symptomatologic unit, as in the classical Galenic definition contained in *De locis affectis*: "Melancholics differ one from the other; all are afraid, scared, blame life, hate the company of other humans, but not all of them wish to die. . . . Hippocrates was therefore right in pinning down all possible symptoms of melancholics to these two: fear and sadness."[22] All later accounts of melancholy, and certainly early modern ones, faithfully repeat this dyad. From a different diagnostic angle, melancholy is also a unit insofar as it is identified symptomatically as a mental affliction without fever, as Galen does, in yet a different work, in his classic definition: "an affliction which affects the mind not without upsetting the spirit and triggering an aversion for things previously loved, but without fever"—a formulation that remains almost unaltered in the majority of early modern accounts.[23]

As part of this cumulative history, melancholy was associated early on with the practice of studying. Rufus's clinical cases abound with references to the nefarious effects of exaggerated mental activity, which supposedly fostered dryness and thereby a prevalence of the black dry humor in the body.[24] The medieval textbook *Regimen sanitatis Salerni*, still in use during the Renaissance, advised avoiding thought and care because they would dry up the body.[25] A crucial document of this interpretation of melancholy's etiology is represented by the pseudo-Aristotelian text of "Book 30" of *Problems*, which maintains that many melancholics are also great minds, generally describable as people who exceed normal limitations, or περιττοί: literally, the ones "beyond," as the Greek prefix περι- indicates.[26] The neutral marker of exceptionality, which in the text of the *Problems* could, in fact, mean exceptional madness as much as exceptional insight, demarcates a space "beyond" that is both doomed and sacred, diseased and gifted. It is this same liminal space that also hosts early modern glass humans.

Following Klibansky, Panofsky, and Saxl's study, the knowledge of an intellectually positive melancholy implied in the pseudo-Aristotelian text was massively reactivated during the Renaissance within the particular context of Italian culture: "only the men of the Quattrocento, with their new conception of humanity, drew from it conclusions amounting to a basic revaluation of the notion of melancholy and to the creation of a modern doctrine of genius."[27] Whether this genealogy is entirely correct or not, it points to the fact that genius was always an aspect of melancholy, as much as "reactivation" can be pinned at any other point in history. As will become clear in what follows, both lines of melancholic culture—that is, the Galenic one of fear and sadness and the pseudo-Aristotelian one of extraordinary capabilities—meet in the figure of early modern glass humans.

Human Pitchers

The figure of glass melancholics circulates, replicates, and undergoes slight variations between the sixteenth and seventeenth centuries. Each piece of documentation is relevant and disclosing some potentially fundamental facets of this historical syndrome.

Before medical texts, the glass delusion is cited in a historical, anecdotal source. In his *Commentarii*, handwritten in the middle of the fifteenth century, Enea Silvio Piccolomini (Pope Pio II) reports that French king Charles VI (also known as Charles "le Fou") suffered from the condition.[28] But the first early modern medical treatise to explicitly refer to the glass delusion is *De habitu et constitutione corporis*, compiled by Dutch physician Levinus Lemnius, and first published in 1561:

One thought his buttocks were made of glass, therefore performed all of his business and his actions standing up, fearing that if he sit down on a chair, his buttocks would have shattered, and the glass shreds would have broken up.[29]

Further accounts of the glass delusion appear quickly thereafter, beginning in 1597, in the treatise on melancholy by the French doctor André Du Laurens:

A great Lord, who thought himself to be glass, and had not his imagination troubled, otherwise than in this one only thing, for he could speak marvelously well of any other thing. He used commonly to sit, and took great delight that his friends should come and see him, but so as that he would desire them, that they would not come near unto him.[30]

In France, the royal physician Louis de Caseneuve, in his didactic compendium from1626 *Hierogliphica et emblemata medica*, mentions as part of the "Phantasmata Melancholicorum," the phantasmatic hallucinations of some melancholics, the case of a Parisian glassmaker, located in Saint Germain, suffering from the delusion.[31]

Despite a surge in studies devoted to the subject of melancholy in the early modern period, also owing to the new economy of circulation of printed texts in Europe, accounts of melancholy do not always refer to the glass delusion, indicating a relative "rarity" of this symptomatology among other possible ones. [32] While appearing in several places, the glass delusion is absent from some highly popular texts, which signals a marginality of the syndrome within the broad melancholic spectrum.[33]

But the anecdotal force of this rare syndrome is such that it is passingly mentioned in Descartes's *Meditations* (1641) as a flagrant and paradigmatic example of madness: "But such people are insane, and I would be thought equally mad if I took anything from them as a model for myself," concludes Descartes after mentioning, among others, those melancholics that believe they are made of glass.[34] Rhetorically hyperbolic, the example owes its efficacy to a visually strong representation of what the loss of rational control over one's mind—but also, crucially, over one's body—might feel like.

Despite the sparsely anecdotal yet hyperbolic power of this figure, however, what counts most is the fact that some people must have indeed fallen ill with this hallucination. Only by recovering the broader etiology and symptomatology ascribed to the melancholic disease can we understand why and thereby locate this figure beyond its occasional apparitions.

The first way to inscribe this syndrome within the melancholic complex is to see it in relation to other delusions—the aforementioned "phantasmata melancholicorum." The largest compendium of melancholy published in the seventeenth century, the vernacular English treatise *Anatomy of Melancholy* by the clergyman Robert Burton, published in 1621, contains three mentions of the glass delusion amid other comparable hallucinatory fantasies.[35] The first appears in the section entitled "Symptoms or Signs in the Mind":

> [Some are afraid] that they are all glass, and therefore will suffer no man to come near them: that they are all cork, as light as feathers; others as heavy as lead; some are afraid their heads will fall off their shoulders, that they have frogs in their bellies.[36]

The second appears in the section "Particular Symptoms from the Influence of Stars: Parts of the Body and Humours":

> Another thinks himself so little, that he can creep into a mouse-hole: one fears heaven will fall on his head: a second is a cock; and such a one, Guianerius saith he saw at Padua, that would clap his hands together and crow. Another thinks he is a nightingale, and therefore sings all the night long; another he is all glass, a pitcher, and will therefore let nobody come near him, and such a one Laurentius gives out upon his credit, that he knew in France.[37]

The third and last mention is found in the section "Symptoms of Head Melancholy":

> As Laurentius supposeth, those feral passions and symptoms of such as think themselves glass, pitchers, feathers, &c., speak strange languages, proceed a calore cerebri (if it be in excess) from the brain's distempered heat.[38]

In these passages, the glass delusion is repeatedly associated with other dysmorphic experiences of body perception.[39] This larger dysmorphic complex appears again and again in early modern texts concerned with manifestations of madness. As Italian author Tommaso Garzoni reports in his 1586 bestseller on mad people, *The Hospital of Incurable Fools* (*L'ospidale de' pazzi incurabili*), there is a melancholic who believes his body is one huge head, one who believes himself a heel, or a cantaloupe in another case, or again a grain of mustard.[40] In all these cases, one's body undergoes a radical transformation leading to some form of impairment.

André Du Laurens, whose anecdote about a great Lord of glass we have seen before, and whom Burton quotes in the passage above (as Laurentius), had included, under the paragraph "Histories of certain melancholic persons, which have had strange imaginations," the case of the "melancholic man which took himself to be a pitcher, and prayed all that came to see him, not to come near unto him, lest they should dash him in pieces."[41] Du Laurens also mentioned, again in plainly anecdotal fashion, the one who believed himself a cock, or to have no head, or to possess a huge nose, or those who believed themselves dead, or even the one who thought he "was nothing."[42] One more glass melancholic who appeared in his treatise imagined "his feet were made of glass, and dares not to walk lest he should have broken them."[43] And the instances of body dysmorphia go on with the case of a baker afraid of being made of butter and with the "silly melancholic" who is afraid to empty his bladder for fear of drowning his town. Compared to these, those who "think themselves kings, emperors, popes, cardinals" would seem to suffer from less unsettling delusions.[44]

A common feature of all these *phantasmata*, or delusional imaginations, is a fantasy of fragmentation, mutilation, or transformation of the body—especially into inanimate objects—in ways that fundamentally alter the location and orientation of the body in space, including bodily contact and exchange with others. The body becomes unable to perform expected movements and functions. By means of this corporeal disempowerment, the social interaction of the sick person is compromised, impeded by fear of touching or by the perceived impossibility of inhabiting the social space.[45] The available significations traditionally imparted to melancholy in the Galenic unit of symptoms of fear and sadness (*metus et moestitia*) find their reconfiguration in the pathological narrative of a body that cannot perform its functions properly because it is afraid of touch, as well as of its own new and unusual proportions: being too big, too small, or even disappearing altogether. Here the embodiment of melancholy means, first and foremost, the bodily reinscription of fear and sadness as received meanings of the disease.

I will return later to the crucial question of embodiment while trying to further analyze the material components of the fantasy of glass humans. For now, what the impairment of the body transforming into glass clearly signals is an interruption in the continuity of the dimension of socially sanctioned normality. This impairment, however, may be also read as a form of resistance through failure to perform, as "glitch" in the programmed system of social efficiency.[46] From this standpoint, the glass melancholics may interpret the function of a posthuman challenge to the humanist status quo that generates and ostracizes them at the same time.

Glass Humans Between Enfermedad *and* Agudeza

Tomás Rodaja, according to Cervantes's narrator, presents a very strange form of madness, or *locura*. In the text, his disorder is also variously defined as *imaginación* or as *extraña imaginación*.[47] Vidriera experiences being made of glass as a condition that is "real and true," *real y verdaderamente*.[48] The assumption of the truth of his predicament allows him to reorganize his world around that fantasy in a perfectly logical way. Tomás's world during the course of his madness is consistently organized around his imagination, down to the most minute details of his material existence: he wears only loose clothing with no shoes, he eats only fruits, he drinks only water collected in his hands from a stream, he walks only in the middle of the street, he sleeps either in open fields or in haylofts, all wrapped in straw; in other words, he does what a real glass man ought to do. Tomás doesn't just believe he is made of glass; he becomes a body of glass to the extent that his imagination is reconfiguring his actual, concrete relation to the world.

If this reorganization of Tomás's existence around his delusion acts first of all as limitation, occasioning a disempowerment of the body within the social space, the story reveals a further set of features attributed to Vidriera's body of glass, which exposes a very different dimension to the glass human's embodiment. Vidriera, through the metamorphosis of his body into glass, becomes a quick-witted and caustic street satirist of sorts. This transformation sustains a long section of the short story that records the dialogues, filled with puns, between Vidriera and the crowd: "He asked people to address him from a distance, and they said that they might ask what questions they liked, because he was a man of glass, not flesh, and since glass is of subtle and delicate matter, the soul works through it with more speed and efficiency than through the material of the normal body, which is heavy and earthy."[49]

Tomás declares himself able to answer with enhanced wisdom *because* he is made of glass. The materiality of glass, "subtle and delicate," replaces the "heavy and earthy" materiality of the flesh. Glass as opposed to flesh, *vidrio* vs. *carne*: the new embodiment sustains a new relationship between soul and body that is productive of new skills and possibilities. How exactly this embodiment plays out in relation to the materiality of glass is a question I will address in more detail in the next section, when exploring early modern glass in its material history. Here, I want to examine more closely what these skills look like in Cervantes's account.

The narrative capitalizes on the state of enhanced efficiency of Vidriera's intellect by creating a subtle play of echoes between the logic governing Vidriera's delusion and the logic governing satirical wit. In other words, Cervantes spells out the rhetorical stakes of Vidriera's delusion to generate narrative entertainment, intertwining the layers of meaning disclosed by the delusion with those accessible to literary form. For example, during his time at court, the glass graduate is interrogated about an array of subjects, which include poetry and poets. After dispelling much of the aura of the poet, he's asked about the reason why poets should be poor. This is an occasion for Vidriera to assert that poets can only be willfully poor, since their poetry instead constantly creates a world of riches:

> He replied that it was because they wished to be, because it was in their power to be rich, if they would only use the chances they had all the time in their hands, or rather, in the hands of their ladies, for these latter were extremely rich, since their hair was of gold, their foreheads of polished silver, their eyes of green emeralds, their teeth of ivory, their lips of coral and their throats of transparent crystal, and they wept liquid pearls. And furthermore, the earth their feet walked upon, however hard and sterile it might be, at once produced jasmines and roses; and their breath was of pure amber, musk and civet; and all these things were signs and marks of their great wealth.[50]

While the passage mobilizes an established repertoire of Petrarchist metaphors (hair of gold, a brow of burnished silver, eyes like green emeralds, teeth of ivory, lips of coral, and a throat of transparent crystal), Vidriera is here displaying a literalist logic (poets should be rich, based on what they do with words) that deftly mirrors the very "literalist" logic of his delusion. While poets may come across in this observation as pedantic fools at best and pathological visionaries at worst, Vidriera comes across, conversely, not just as a pathological character but as a poet himself. But the bewildering equivalence between poetry and delusion holds up to a certain point: poets just pretend to predicate the reality of their metaphors, while Vidriera is certain of his world, *real y verdaderamente.* While the assertive predication of truth might be logically equivalent, the commitment to truth is practically very different in the two cases. And for Vidriera, there is little doubt that he is not faking it, as is proven by his open acknowledgment that there was a time when he wasn't made of glass ("una vez, cuando no era de vidrio") and a time, unlike now, when he was made of flesh: "cuando yo era hombre de carne, y no de vidrio, como agora soy."[51]

Yet it is Vidriera's literalist attitude that propels the satirical punch of his puns. When a passerby asks him "what he should do to go off with a job" (*salir con una comisión*), he replies: "Get on your horse along with the person who has the job, and accompany him as he leaves the town, and you will go off with it" (*saldrás con ella*).[52] The graduate's whole art of puns is devoted to homophones, amphiboles, and the art of *double entendre*, where the logic of the *word* becomes the logic of the *world*:

- Mr Glass, a money-changer ["banco"] who was condemned to be hanged died in the prison tonight.

[. . .]

- He did well to hurry up and die, before the hangman sat on his bench ["banco"].[53]

In another pun, addressed to a small group of Genoese citizens that had talked to him:

- Come over here, Mr Glass, and tell us a story ["cuento"].
- I'd better not, you might transfer it to Genoa.[54]

It is not that Vidriera cannot distinguish metaphors from literal meanings: "Although I am made of glass, I am not so weak as to allow myself to be borne along by the vulgar mob."[55] Rather, he is able to produce literal meanings from metaphors, to literalize figural expressions. Hallucinatory and logical at the same time, the *licenciado* speaks and jokes the same way he lives in his glass body: his words become things; *phantasmata* become beings.

As delusional as he is, Vidriera is sharper than the crowd; and unlike the crowd, by his own estimation, he is not "deceived" (*engañado*). The opposition *vidrio/carne* provided by Vidriera himself as the key to his intellectual skills is a crucial point for the story, as it discloses the properly symbolic dimensions of the glass delusion as the imagination of an alternative, enhanced form of embodiment. The alleged property of glass to be a lighter, more agile vessel for the soul, as opposed to the fleshy cage of the body—the latter a *topos* of both scriptural and Platonic derivation, as we'll see—is not something that medical accounts of the disease ever even dwell on. In other words, the suggestion we find here allows us to establish a semiology for the glass delusion that nonliterary accounts are reticent about, be it because they already imply it or because they do not consider it relevant.

If the materiality of the metamorphosis into glass is certainly impairing and disempowering, articulating the embodiment of fear, it also appears to be a vehicle for a different, privileged relation between body and soul, which adds and specifies an alternative phenomenology of the embodiment in glass. Because of the latter dimension, in the course of Cervantes's story, people start looking up to Vidriera to receive insights or to just be astounded by his sharp wit:

Some people, wanting to see whether what he said was true, asked him many and difficult questions, to which he replied spontaneously, with great sharpness of wit [*con grandísima agudeza de ingenio*], and this caused great wonder to the most learned people of the university and to the teachers of medicine and philosophy, when they saw that a man who was the victim of such an extraordinary delusion [*extraordinaria locura*] as to think that he was made of glass, had within him such an acute intelligence [*tan grande entendimiento*] that he could reply to any question appropriately and shrewdly [*con propiedad y agudeza*].[56]

The surprise of the academics in the quote is, however, not entirely justified. According to the early modern conception of the structure of the mind, articulated into faculties and functions, imagination could be affected by some disorder even when rational capabilities remained intact.[57] According to Renaissance theories of the mind, which were very much a continuation of medieval Aristotelianism such as Aquinas's, the mind was a structure of faculties interacting with different faculties of the soul: vegetative, sensitive, and rational.[58] The sensitive faculty was mapped onto the three brain ventricles and articulated into the three main faculties of imagination (*vis imaginativa*), cognition (*vis aestimativa* or *cogitativa*), and memory (*vis memorativa*).[59] Most Renaissance accounts of the mind are indeed based on this topography,[60] and in the context of melancholy, imagination was mostly indicated as the faculty affected by the melancholic humor, also because of its supposed location in the front ventricle of the brain.[61] Conversely, it was mostly the imagination that was held responsible for the alterations of judgment in the melancholic person. Technically speaking, it was thus perfectly possible, in Vidriera's state of mental illness, that the strangest imagination (*locura, imaginación, enfermedad*) coexisted with the sharpest intellect (*agudeza de ingenio*).[62]

While this topography of the mind alone would be able to explain, within the framework of knowledge available to Cervantes, Tomás's coextensive *enfermedad* and *agudeza*, the story provides further valuable hints, otherwise missing from the medical accounts, as to why a glass human should be able to speak better and more incisively about any subject. The text provides an explanation that prompts us to look at the specific material inscription of the delusion: "Since glass is of subtle and delicate matter [*de materia sutil y delicada*], the soul works [*obraba*] through it with more speed and efficiency [*con más prontitud y eficacia*] than through the material of the normal body, which is heavy and earthy [*pesada y terrestre*]."[63]

The new body of Vidriera endows him with new capabilities. In the next paragraphs, I will unpack his glass corporeality to identify the historically specific connection between the materiality of glass and the glass delusion. The material qualities of glass, albeit partly overlapping with contemporary ones, must be understood in their historical context to avoid the common pitfall of reading the short story under the contemporary metaphoric of glass, through a set of figurative meanings that are not the product of the culture that generated that figure.[64] In fact, by applying a metaphorical network that is not the product of the very discourse that generated the glass delusion, most critical literature on Cervantes's "Vidriera" seems to have failed to recognize precisely those dimensions that I have indicated as crucial to it—namely, the melancholic inscription of the graduate's madness and the embodied condition of Vidriera's madness. It is especially the embodied experience of Vidriera's *locura* that seems to disappear from those critical accounts keen on metaphorizing this syndrome[65]—something puzzling, considering that the text dwells quite significantly on Vidriera's fear of being broken, almost providing a phenomenology of physical pain under the conditions of breakability.

Recall the description of Vidriera's reactions when people come close to him: "The unfortunate young man imagined that he was all made of glass, and in this delusion, whenever anyone approached him, he would shriek, begging and pleading with coherent words and arguments, for people not to come near, because they would break him, because really and truly he was not as other men—he was made of glass from head to foot."[66] Further passages insist on the aspect of the fear of breaking: "when it thundered, he shook like a leaf, and went out into the countryside";[67] when street boys throw rags and stones at him for mockery and provocation, "he shrieked so loudly and was so alarmed that he caused the men present to shout at the boys."[68] One day, we are told, "a wasp stung him on the neck, and he dared not shake it off in case he broke, and yet he complained all the same."[69] Vidriera's reaction to pain is primarily informed by his greater fear of breaking, which forces him to endure the sting of the wasp. But the logic governing his pain is soon questioned: "Someone asked him how it was that he could feel the wasp, if his body was made of glass? To which he replied that that wasp must have been like a slanderer, since the tongues and beaks of slanderers were enough to crumble a body of bronze, let alone of glass."[70] Vidriera shows here, once more, his shrewd talent for literalizing metaphors to satirical effect. But his shrewdness in this context is used to dismiss the accusation that he is being illogical—that he, in fact, should be insensitive to pain if he is made of glass. Ultimately, what Vidriera argues for here is the legitimacy of his pain based on his being made of glass—the legitimacy of his being a body, though one made of glass. Glass can be affected by pain, in his world of meanings, as the metaphor allows bronze to get crumbled by the beaks and tongues of slanderers.[71] By assigning priority to the dimension of embodiment in Vidriera's disease, we cannot help but see that his being of glass is not just a metaphor for various possible moral or cognitive meanings. The body of Vidriera, or of those glass humans coming up in the medical treatises of the early modern European era, is not a metaphor, as a body can never be. Here, a specific anachronistic employment of philosophical tools may be allowed, especially in light of the fact that the glass body of the melancholic challenges the very historical episteme that has produced it. In his classic phenomenological definition of bodily experience, Merleau-Ponty defines the body as an "expressive unity that we can only learn to know by taking it up."[72] To take up Vidriera's body, we have to explore what glass was and what it looked like back in the time when Vidriera meandered the streets of Salamanca.

Fragility and Transparency: The Materiality of Glass

The glass that early modern glass humans believed they were made of was not exactly a new material, yet something about it was definitely new, as we will see presently.[73] In fact, before early modern texts would start speaking of glass humans, a similar delusion adhered to a different cluster of materials and objects, among which the clay vessel was the most relevant. Galen mentions, among melancholic delusions, those melancholics who are afraid of being made of clay.[74] As Harald Weinrich has pointed out, in the Renaissance the body was regarded as a vessel of the soul, consistent with a biblical and Pauline topos.[75] The breakability of the clay plays out, in this textual tradition, in different constellations of meaning. The biblical Ecclesiastes evokes a broken pitcher as the figure of disintegration and death,[76] while the books of Job and Isaiah emphasize the material creation of the human in clay by the hands of God.[77] The Pauline reelaboration of the biblical topos in Corinthians presents an ambivalent allegorical materiality: "But we have this treasure in clay jars, so that it may be made clear that this extraordinary power belongs to God and does not come from us. We are afflicted in every way, but not crushed; perplexed, but not driven to despair; persecuted, but not forsaken; struck down, but not destroyed; always carrying in the body the death of Jesus, so that the life of Jesus may also be made visible in our bodies."[78] The Pauline clay jar, despite its fragility, does not break but is in constant peril. Even more important, however, it contains a treasure, a precious substance that should not be spilled or wasted.

The vessel-shaped object deeply informs the early modern glass delusion: in the variant we have seen repeatedly, those afflicted believe themselves to be vases and vessels of various kinds (often pitchers). Cervantes writes that Vidriera asked for some straw to cover "that fragile vessel [*vaso*] of his body."[79] The motif of the vessel supplements the idea of fragility with the idea of a valuable content that risks going to waste. The nature of this content is, however, also variable. Further objects share in the symbolic network attached to the glass delusion. The word *vidriera*, for example, in the referential system of Cervantes's times, indicated not just a windowpane—another object implying specific functions and meanings, some of which I will discuss later on—but also a glass urinal—that is, the bottle where urine was kept for diagnostic but also "clairvoyant" purposes, since the flask could be used to determine a person's state of health *in absentia*.[80]

Glass had certainly been around long enough to replace clay in the human imagination, yet it was only an early modern substitution. This fact cannot be underestimated in looking at the historical formation of the glass delusion, and I believe it urges a reading centered on the novelty of glass production that chronologically coincided with the telling of Vidriera's fictional adventures.

Even before looking at early modern glass, however, let it be clear that glass had always been brittle and had, therefore, always been employed figuratively to designate fragilities. To be a "man of glass" was, for example, an idiom in medieval Italy by at least 1363, when Giovanni Boccaccio was called a "man of glass" by a detractor pointing out his excessive sensitivity and tendency to take offense. In an epistolary response to his accuser, Boccaccio, who felt seriously insulted by the epithet of *uomo di vetro*, replied: "We are all glass men, subjected to innumerable dangers; the slightest touch would break us and we would return to nothing."[81]

The fragility of early modern glass objects had spun a further line of analogical thinking tinged with Christian overtones. Vannoccio Biringuccio, a metallurgist from Siena who wrote about metalworking and glass in his *De la pirotechnia* (1540), says about glass: "But considering its brief and short life due to its fragility, one cannot and should not bestow much love upon it; and as an example, one should use it and keep it before one's mind as a reminder of the life of man and of the things of the world, which are fleeting and fragile, even though they may be beautiful."[82]

The glass-like fragility of life is thematized in the aforementioned medical work by Louis de Caseneuve, who provided the anecdote of a glassmaker from Paris Saint Germain. At the beginning of his exposition, conceived as a comprehensive and didactic visualization of fundamental facts, he presents a paradigmatic analogy between human life and glass. The emblem of a pyramid is explained as the pyramid of life—a pyramid made of glass:

> Life is a pyramid made of glass; for life is like glass—fragile, indeed even more fragile than that. . . . Truly, human life is far more fragile than glass, and so much more so as things that break spontaneously compared to those broken only by an external force. . . . Beyond the fragility of glass, its composition must also be considered and related to life. Glass consists of two substances: one is vegetative, which is spiritual and volatile, from which fusion is possible; the other is mineral, which is corporeal and fixed.[83]

Further on, the sentient being is even said to look like a glass-blown object: "veluti operis vitriari." The passage establishes a foundational connection of life to the material potential of glass, insisting on fragility but also on the fascinating metamorphic, plastic qualities of the material.[84]

While brittleness had always characterized glass, the history of its transparency is distinct and has developed at its own pace.[85] I contend that the transparency of early modern glass humans is, in fact, perfected and active in the delusion—but not necessarily in the way we might expect (the see-through effect of transparent glass). The melancholic glass humans of medicine and literature we are looking at are perhaps not so much concerned with the visibility of their inner dimension—a concern that will, however, become essential in the unfolding history I am telling in this book—but with the way they can see through the world.

Gill Speak has retrieved the figure of a glass man in a British play from the early seventeenth century, Thomas Tomkis's *Lingua* (1607), in which the transparency of the glass melancholic seems more or less explicitly indicated.[86] Enacting the battle between language and the five senses, the play represents Tactus (the sense of touch) as a glass man, portrayed in conversation with Olfactus (the sense of smell), to whom he admits his delusion, all with obvious comical implications. The transparency clue is provided by Tactus, who tells us that he found he was made of glass when exposed to direct sunlight that filtered through his fingers, and then discovered a window in his chest "through which I plainly did perceive my heart." His interlocutor, Olfactus, thus exclaims, mockingly: "Momus himself can find no fault with thee / Thou'st make a passing live Anatomie." The passage, which for Speak proves the ascription of transparency to the glass melancholics, is particularly interesting because it ties the figure of the glass melancholic back to the fantasy of Momus, which is crucial to the arc of the story I tell about transparent humans in this book. This document, then, attests to the presence, in the glass delusion, of an element of autoscopic self-revelation and exposure as is the myth of Momus—a myth of self-revelation, unthwarted communication, and mechanic perfection. But as much as this represents a salient clue in unraveling the still enigmatic existence of early modern glass humans, the function of transparency that the early modern figures employ and capitalize on is a different one—in fact, its inverse.

In my research into the early modern material history of glass, I have come across sources that tie the glass syndrome even closer to the materiality of glass. I propose thereby a new material history for the glass delusion, which makes it possible to reinscribe it within the melancholy complex—not just as one of its marginal forms but as a particularly condensed figure (albeit minor) of melancholy's meanings within European early modern culture.

The Man Who Became a Bottle: Glassblowing and Murano

Italian Renaissance bestselling author Tommaso Garzoni, mentioned above as the author of *The Hospital of Incurable Fools*, wrote another encyclopedic and entertaining compendium of characters—or, more accurately, personality traits—which he defined as types of "brains." The work, entitled *The Theatre of the Many Worldly Brains* (*Il teatro de' vari e diversi cervelli mondani*) was composed in Italian vernacular in 1583. A section of this partly whimsical, partly sententious work on human types is devoted to "melancholic and wild brains" (*de' cervellazzi menanconici e salvatici*). Among the many entertaining anecdotes, dressed in a colorful popular style, Garzoni tells the story of a man who, "thinking he had become a piece of glass, went to Murano and threw himself in a furnace to be blown into the shape of a bottle" (*in foggia d'una inghistara*).[87] This anecdote discloses the kind of interaction that took place between early modern melancholy and early modern history of glass in some fundamental ways.

The man Garzoni is writing about not only thinks he has become glass, but he intends to be further transformed, through the art of glassblowing, into a particular kind of glass vessel, an *inghistara*, a bottle endowed with a large round base and an elongated neck. This double transformation is an indication of the deeply metamorphic, and not just dysmorphic, pattern that structurally organizes his imagination.[88] This extensive metamorphic fantasy is supported by the technical possibilities implemented in the craft of glassblowing that was developing on the island of Murano since the fifteenth century with particular force and that peaked in the seventeenth century.[89] Looking at artifacts or designs from that time may disclose the imaginative horizon available to the material culture of glass at that particular moment.

In Giovanni Maggi's *Bichierografia* from 1604, a compilation of shapes of bottles and vessels of blown glass showcases the pliable versatility of the glassblowing techniques in achieving any kind of desired shape. Maggi's work has to be understood as a visual catalog of designs, possible objects at the crossroads between the reality of artifacts and the fantasy about their possible shapes. In these drawings, precious items from aristocratic collections are juxtaposed with fantastic items imagined by the author, without marking distinctions or boundaries between the two kinds of objects, the real and the imaginary (fig. 1 and fig. 2; in fig. 3, we can see the correspondence between a Muranese vase and one of Maggi's designs, both slightly anthropomorphic). This fluid border between reality and fantasy was precisely a product of the flourishing of glass craftmanship starting with the fifteenth century and may help explain more than anything else the rise of the fantasy of glass humans. In the *Bichierografia*, very elaborated shapes achieved by skilled craftsmen are mixed with imaginary human-shaped vessels (or even glass humans). The human figure appears, pointedly in the selection reproduced here, as a liminal-object: a vessel that served no purpose other than (possibly) decoration. It can be read in two directions: as a glass vessel in the shape of a human body or as a human body made of glass. Maggi's repertoire as a collection of imagined objects defines the desirability of glass embodi-

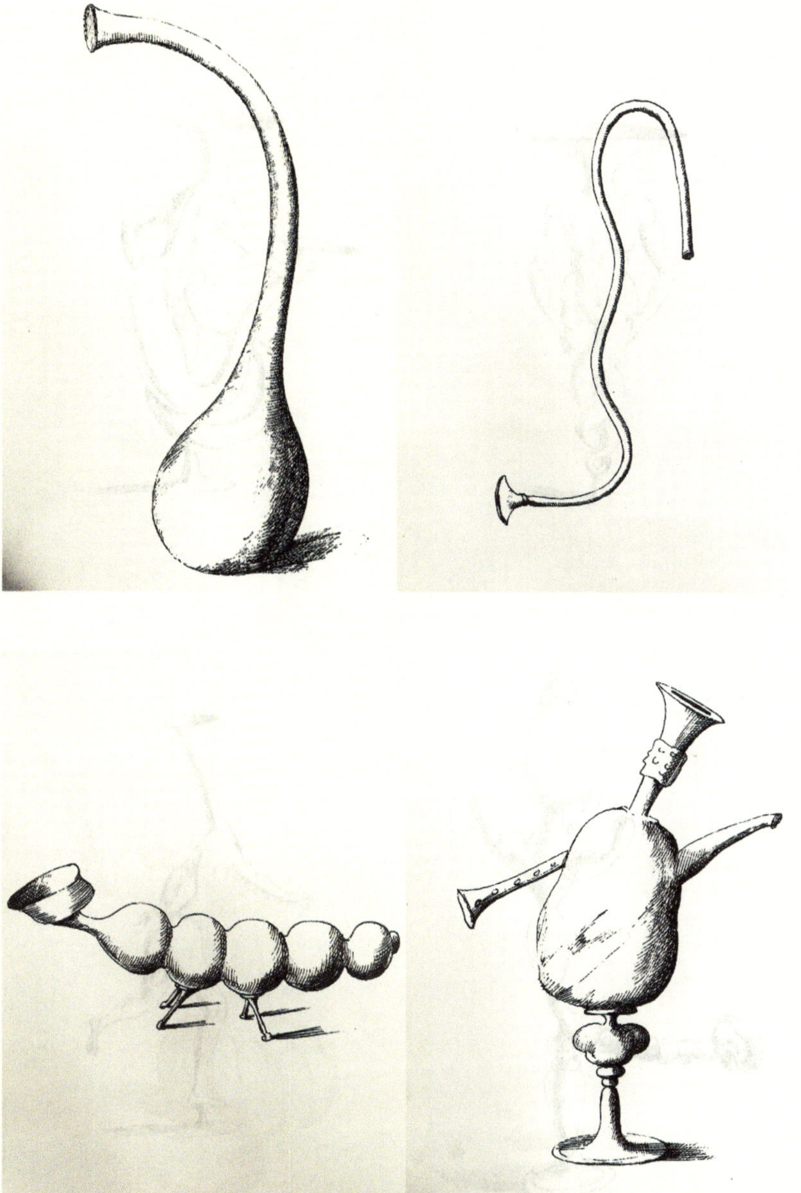

FIGURE 1. Selected images from Giovanni Maggi, *Bichierografia* (1604). Facsimile edition, 4 vols., ed. Paola Barocchi (Studio Edizioni Scelte, 1977).

FIGURE 2. Selected images from Giovanni Maggi, *Bichierografia* (1604). Facsimile edition, 4 vols., ed. Paola Barocchi (Studio Edizioni Scelte, 1977).

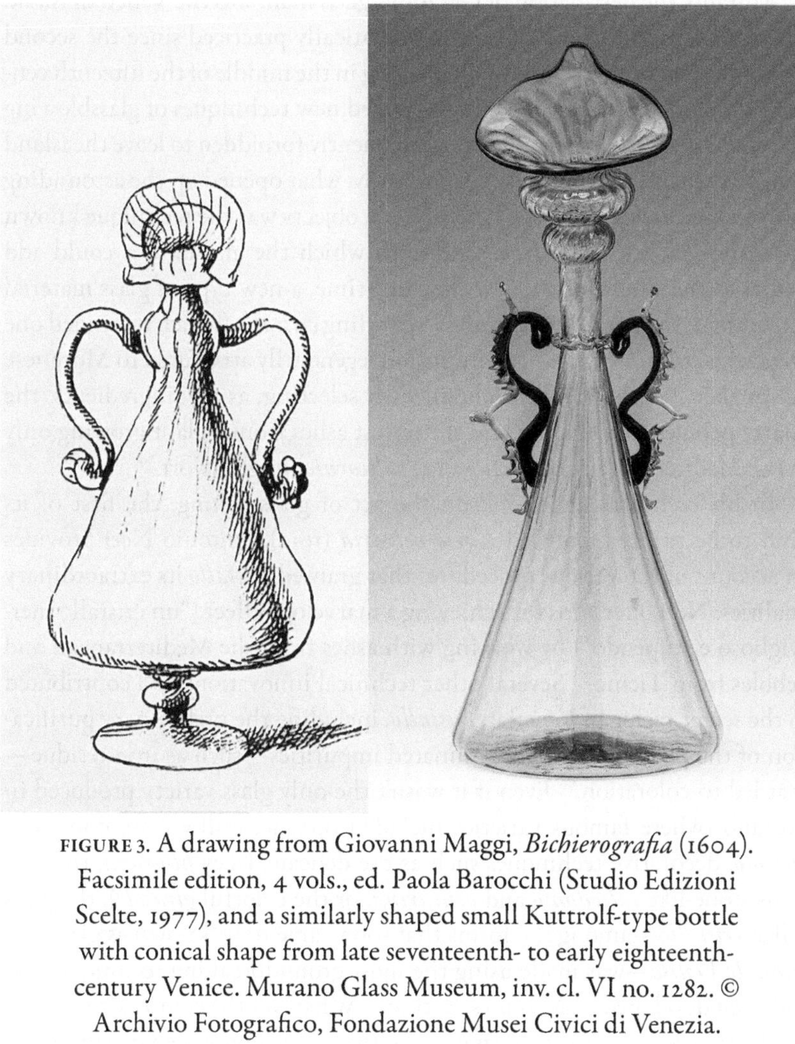

FIGURE 3. A drawing from Giovanni Maggi, *Bichierografia* (1604). Facsimile edition, 4 vols., ed. Paola Barocchi (Studio Edizioni Scelte, 1977), and a similarly shaped small Kuttrolf-type bottle with conical shape from late seventeenth- to early eighteenth-century Venice. Murano Glass Museum, inv. cl. VI no. 1282. © Archivio Fotografico, Fondazione Musei Civici di Venezia.

ment in relation to the technical and aesthetic possibilities allowed by the new developments in glass craftsmanship. It is an invaluable document for understanding the imaginative and material knot of associations implied in the figure we are reconstructing.[90]

Murano, the destination of Garzoni's glass man, was the Venetian island where glass production had been systematically practiced since the second half of the thirteenth century.[91] Beginning in the middle of the fifteenth century, the skilled artisans of Murano created new techniques of glassblowing and a new glass material and were subsequently forbidden to leave the island and share their secret recipes.[92] Technically, what opened up the astounding morphological possibilities of glass-blown objects was the technique known as "mano volante"—the free hand with which the glassblower could add shapes to the simple mold.[93] At the same time, a new type of glass material was crafted: the amazingly colorless, sparkling *cristallo* (which reminded one of quartz, *cristallo di rocca*), an invention legendarily attributed to Muranese glassmaker Angelo Barovier, obtained by selecting, as raw ingredients, the quartz pebbles from Ticino and the purest ashes from a plant growing only in the Mediterranean zone, known as *salicornia*, or glasswort.[94]

In his technical handbook on the art of glassmaking, the first of its kind to be printed, entitled *L'arte vetraria* (1612), Antonio Neri provides an account of the various procedures that granted *cristallo* its extraordinary qualities. Neri offers tips for achieving a marvelous effect ("un cristallo meraviglioso e stupendo") by working with ashes from the Mediterranean and pebbles from Ticino.[95] Several other technical innovations had contributed to the secret recipe of Venetian *cristallo*, including the preliminary purification of the plant ash, which eliminated impurities—such as iron residue—that led to coloration.[96] Even if it wasn't the only glass variety produced in Murano (where famous varieties included opaque, milky glass known as *lattimo*, decorative techniques such as the delicate, lacey *filigrana*, the precious stone-like *calcedonio* and *venturina*, or the colorful *girasole*), the glass called *cristallo*—among the forms that soon came to be known as glass *à la façon de Venise*—was made using the most groundbreaking technique that had been developed by Muranese artisans. Whatever the complexities of the production process, one thing is certain: the *cristallo* produced in Murano at that time was the purest, most transparent type of glass ever produced up to that time.[97] Its unprecedented thinness endowed it with an "incomparable elegance of line" and, inevitably, with extreme fragility.[98]

Garzoni's anecdote about the glass man who wanted to be melted in a glass furnace shows us how the new objects produced in those furnaces, with their outstanding qualities of thinness and transparency, left a profound mark on the imaginations of those who witnessed their diffusion. When in 1626 French doctor Louis de Caseneuve reported, in the anecdote quoted above, on how a Parisian glassmaker had become convinced his buttocks were made of glass, he provided us with something more than just another anecdotical variation. By mentioning a man who, working with glass, started to conceive his body as made from glass, he was hinting at the most direct possible link between the circumstances of glass manufacturing and those of the imagination of the glass men. The novelty of glassblowing's endless possibilities of shaping objects nourished the fantasy of (or even the drive for) metamorphic processes, as it is theatrically illustrated in the engraving of a furnace by Giuseppe Maria Mitelli from the end of the seventeenth century (fig. 4). A group of men observes ("Io osservo," says one) a glassblower whose blowpipe has just produced a figurine of glass who speaks: "I am made of glass, and I am made of blown air" (*Fatta di vetro io son, e son di soffio*). On

FIGURE 4. Giuseppe Maria Mitelli, *Fornace da vetri* (Glass Furnace) (1698). Engraving. British Museum, London. Courtesy of the Trustees of the British Museum.

the left, the skeptical observers intervene: "It is difficult to make!" (*È difficile a farla!*)—says one—"But it is easy to break it" (*Ma è facile a romperla*)—replies the other. One more wise comment sums it all up: "We will see when the work is done" (*Si vedrà nel fin dell'opra*). The wonder of the glassblowing process emerges from the reactions of the beholders, but its counterpart is the wonder of the talking glass figure, created by the blowpipe and the breath (*soffio*) of the glassblower.

Not only did the blowing into the glass to produce shapes resonate with the biblical pneumatic vocabulary of creation, but the gestures implied in the work of the glassblower (*soffiatore*) at the furnaces involved a specific bodily engagement in the productive movements that would only further enhance the idea of the furnace as a location for radical material transformation and the melting of the borders of the human shape.[99] The transformative, potentially mysterious, environment of the furnace is indeed thematized in John Locke's empiricist manifesto when, in *An Essay Concerning Human Understanding* (1690), he writes, echoing (and thereby criticizing) Descartes's apparent eschewing of madness in favor of rationalist dreams: "And if our dreamer pleases to try whether the glowing heat of a glass furnace be barely a wandering imagination in a drowsy man's fancy, by putting his hand into it, he may perhaps be awakened into a certainty, greater than he could wish, that it is something more than bare imagination."[100] The glass furnace was very much a real setting of incredible transformations and possibilities, so much so that the madness of glass humans was perhaps less a fancy than a reasonable thought experiment.

Not just the material and special circumstances of working with glass expose glassmakers to become protagonists of such metamorphic fantasies, but their peculiar social status as craftsmen seems to reflect their intimate symbiosis with the material they can so thoroughly shape at their own will to produce objects of use, as well as of beauty, ready to be contemplated and purchased by wealthy customers. French engraver Nicolas de Larmessin, within a series of images dedicated to professions and stylized as figures of Arcimboldi, bodily hybrids of items, visualized this symbiosis in the shape of a glassmaker (and glass vendor) made of his glass products (fig. 5). Some decades later, German engraver Martin Engelbrecht delivered us a similar illustration of what a glassmaker might look like: the body of the glassmaker (in this case of his wife) has again morphed into an assemblage of glass items (fig. 6). This emphatic metamorphic representation might amplify the ex-

FIGURE 5. Nicolas de Larmessin, *Habit de vitrier* (1680–1700). Engraving. Rakow Research Library, Corning Museum of Glass, Corning, New York (CMGL 137235). Courtesy of the Rakow Research Library.

FIGURE 6. Martin Engelbrecht, *Une verrière / Eine Glasmacherin* (1730). Engraving. Rakow Research Library, Corning Museum of Glass, Corning, New York (CMGL 129856). Courtesy of the Rakow Research Library.

ceptionality of the social stand of European glassmakers, transnational migrants, privileged and proud outsiders in the wide world of craftmanship.[101]

The passage from the clay body (*testaceum*–ὀστρακοῦς) of the Galenic tradition to the glass body, which—based on the presence of regular glass—could have occurred earlier than it actually did (since glass was available long before), is therefore not to be understood merely as a substitution of materials that are just typically fragile. The introduction of glass didn't just provide a new raw material to a preexisting pattern of melancholic imagination that had, until then, been relying on clay; rather, it transformed that pattern from within in two crucial ways.

First, as we have seen in Garzoni's anecdote, the body of the glass melancholic (not unlike that of clay) is breakable, extremely fragile, but—and this is substantially new—it is also pliable: susceptible to multiple changes of form. This element modifies the fear of traditional melancholic delusions by intersecting with the material circumstances of glassmaking—namely, the transformative, creative, and body-intensive process of glassblowing. A look at the astonishing iconographical repertoire of Maggi's *Bichierografia* leaves no doubt as to the imaginative stimuli that sprang from the craft of Muranese glassblowing. Inserting this transformative pattern into the established forms of the melancholic syndrome meant adding a further declension of *metus et moestitia* to the dysmorphic enactment of fear exhibited in many delusions—namely, the fear of changing shape, of lacking stability, of being an object that can morph into a human as well as a human that can morph into an object, like one of those liminal prototypes from the *Bichierografia*. In this new reconfiguration of a classic melancholic delusion, the emphasis shifts to the very instability of the melancholic condition, in which the disruption of "normal" representations doesn't settle necessarily in a deviant but at least stable state; it instead keeps shifting.

A second, crucial way in which the melancholic syndrome of the glass human modifies (while partly reinforcing) previous patterns of the disease is foregrounded by the special thinness of *cristallo*. This property, once related to the melancholic body, adds to the fragility of common glass (shared by clay) a novel subtility and delicacy. Like those precious objects portrayed in Maggi's portfolio, the melancholics who transform into glass objects reclaim for themselves a higher market value. This economical accretion is most powerful because it meets the traditionally established structure of melancholy in ways that will now become clear.

The "subtility" of glass—its extraordinary thinness—is the astounding result of a treasured and proprietary craftsmanship and thus a token of its exceptionality, the same way that the melancholic disorder entails some kind of extraordinary capability. Early modern glass intercepted the melancholic complex where it signaled the melancholic's exceptionality, thereby providing the glass delusion with a paradigmatic function in representing early modern melancholy more at large.

The economic dimension sustaining the surplus of value of a glass body is only hinted at in Cervantes's story. Once Vidriera's fame has spread around the country, for example, he receives an invitation to court, and for the occasion, he is wrapped and packed in straw and transported like those precious crystal objects that were much sought after within the aristocratic circles of Castilla and Catalonia in the sixteenth and seventeenth centuries. On the one hand, in the sixteenth century, the first tours of Murano had begun, in the course of which wealthy travelers were brought to Murano to see the furnaces and purchase objects to bring back home.[102] On the other hand, the Iberian peninsula became, later in the seventeenth century, one of the major centers of production of glass *à la façon de Venise* outside of Murano, and many Muranese emigrants had settled in Spain (as already mentioned, the emigration of glassblowers was not allowed in the Venetian Republic).[103] Only later, starting in the eighteenth century, glass would become a material more largely accessible as a consumer good not only limited to wealthy aristocratic milieus.[104]

The greater value of a glass body extends, however, beyond the economic laws of the market. There is an intrinsic value to the melancholic's glass body that resides in an enhanced organization of the relation between body and soul. It is there that we have to look in order to ultimately and completely "take up" the bodies of early modern glass humans and understand their complex negotiations; that is, we need to look into the philosophical implications inherent in the new and extraordinary materiality of glass.

Looking Through a Window: Early Modern Mind/Body Relations

As we discussed above, Cervantes recounts how Vidriera had become so acute in his intelligence and wit exactly when his body had become glass: "He asked people to address him from a distance, and they said that they might ask what questions they liked, because he was a man of glass, not flesh, and since glass is of subtle and delicate matter [*de materia sutil y delicada*], the soul works through it with more speed and efficiency [*con más prontitud y eficacia*] than through the material of the normal body, which is heavy and earthy."[105] Cervantes's indication that the soul (*alma*) contained in a glass body is quicker and more efficient than a soul in a body of flesh is sustained by two conditions of very different kind: first, that the glass being produced is the subtle and transparent *cristallo* (material premise) and, second, that the soul is being conceived—according to a biblical and then Neoplatonic tradition—as "inhabiting" the body (philosophical premise). While the first premise has been expounded in the previous section, the latter still needs some elucidation.

The idea of the soul inhabiting the body is long-lived but prominently revived in early modern Neoplatonic thought. Platonism and Aristotelianism had variously described and conceptualized the instrumental nature of the body in relation to the soul, implementing a semantic field of items and tools to figuratively describe this relation.[106] Platonism notoriously operated along a mind/body dualism that relied on the representation of the body as hindrance to the soul, expounding multiple semantic fields signifying obstruction.[107] The goal for the soul, in the *Phaedo*, is to be "freed from the body and from fetters,"[108] which are at times pictured as prison bars when not as other forms of bondage indicating that the soul is kept prisoner by the body.[109] In *Phaedrus*, this bond is even formulated as the captivity in a peculiar cage that is, in fact, a house: the oyster's shell.[110]

Christian Neoplatonism had introduced a further semantic field—namely, that of dwelling—to describe this relation. In the *Confessions*, Augustine elaborated a complex semantics of displacement and spiritual exile, making ample use of the trope "house of the soul."[111]

It is precisely this trope that was expounded in quite an original manner by Juan Luis Vives (1492–1540), a cosmopolitan humanist native of Valencia and son of Spanish Jewish converts to Christianity, in a work that Cervantes probably knew: *De anima et vita* (1538).[112] Here, Vives argued that the soul inhabits the body: "the soul dwells [*habitat anima*] in its abode as if within rooms, with all their objects and domestic tools."[113] That is, the soul inhabits the body as a well-furnished apartment—the instrumental nature of the body retained by the image of the tools available in the rooms. The semantics of spiritual interiority conceived of as domestic space intensifies as Vives goes on distinguishing the various functions of the soul. While sharpening, it also contracts into a much more negatively connotated experience of dwelling—that of incarceration. In Vives's account, the rational soul—already for Aristotle its distinctively human, highest expression—is governed by the mind (*mens*), a faculty that relies on knowledge provided by the sensitive soul and the aid of two further faculties: will and memory.[114] The body encircles the mind like a dark prison (*obscuro carcere*), a thick fog (*corporis caligo*), or shadows (*obumbrante corpore*).[115] The mind, the highest tool of human cognition and our species' link to God, constantly needs to resist the constraints of the body. Vives evokes several images to represent the mind's struggles, such as sliding down a slope, being dragged by the current of a torrent, and even wearing colored glasses that provide a deceptive vision of reality.[116] But the most articulate image to capture the inescapable struggles of the mind is one that is particularly significant in our investigation in explaining how the glass delusion of Vidriera allows his *ingenio* to become so sharp and sophisticated.

Vives describes the mind being encased in the body as a man is locked in a cell with only one glass window:

> The mind is in fact locked in this body as somebody is locked in a room and has no other window but a glass one; and cannot see anything but what the glass allows him to see, that is, more clearly if the glass is clean and transparent, much less if the glass is covered with dirt or dust. The mind is like the wiseman in the home of the unwise; of course it depends on the senses, as it is guided by them. But as a man who is locked in a room, who doesn't ignore the existence of the glass before him, through which he might see more or less clearly and openly, so is the mind in relation to the body: though guided by the senses, it corrects them.[117]

This image is accompanied by preceding and subsequent ones: the imagination (one of the faculties of the sensitive soul) is said to operate as "per vitrum," through a window,[118] and the *ratio* is said to be enclosed by the senses as if by a room: "sicut qui clausus est in cubiculo."[119]

Amid disruptions and distractions, the mind keeps inhabiting the limited, closed, and even dusty room of the body, which Vives later even compares to a messy workshop (*officina perturbata*),[120] where tools have become unusable and distracting—unavoidably evoking the melancholic iconography of desuetude that suffuses Albrecht Dürer's *Melancholia I*.[121]

In the allegorical image produced to describe this cognitive as well as experiential situation, Vives claims that the clearer the glass is, the better the mind can see; the dirtier and opaquer the window, the less the mind is able to see through it. The glass window (*vitrea fenestra*) corresponds to the sensory apparatus of the body; a clearer, lighter, unimpeded body will allow a clearer, more effective guidance on the part of the mind.

Leaving aside this semantics of habitation (but providing further support for my claim), later in the same treatise, Vives characterizes the reasoning of the melancholic as exceptionally agile: "if black bile is mixed with subtle and clear spirits, it provides agility of reasoning, of judgment, of prudence and wisdom; those who can go deep create and invent many excellent things."[122] The outcome of this agility (*dexteritas*) and the depth of its reach (*descendunt alte*) are the excellent skills and achievements of melancholic minds.

The early modern melancholic syndrome of the glass delusion rests therefore as much on the material advancements in the enhancement of the transparency of glass (with the corresponding accretion of glass's market value) as on a deep-seated philosophical understanding of the relationship between soul and body, in which the body acts as a limitation on the operations of the soul, the human link to the divine.

Against this background, Vidriera's body seems to offer a solution to the problem of the mind held captive within the body. Among the variations of the glass delusion we have reviewed so far, the imagination of Vidriera has picked precisely a window to encase his soul, just like Vives's captive. But Vidriera's embodiment in a glass that, we have to assume, is exceptionally transparent and thin, is a chance to enhance rational faculties—and it works. Vidriera thus perfectly embodies the type of the *perittós*, the exceptional melancholic of Book 30 of *Problems*. Vidriera, as we have seen, is endowed with rare *ingenio* and extreme *agudeza*. The tradition of the exceptional melancholic joins forces with Vives's figural metaphysics and the technical innovations of glassmaking to present the prototype for an enhanced type of human being, thereby placing the glass melancholic both within and beyond the melancholic complex.

Solutions Made of Glass

Vidriera is not just a paradigmatic embodiment of the duplicity of melancholy between disease and skill, impairment and endowment (following Marsilio Ficino's formula "melancholia est duplex").[123] By embodying an enhanced though deviant human prototype, Vidriera—together with the other glass humans of early modern texts—both epitomizes and defies humanist assumptions of the human, performing at the core of humanism what Rosi Braidotti has called a "posthuman convergence."[124] In this regard, the approach to early modern posthumanism in terms of mere anticipators and "predecessors to the contemporary posthuman subject"[125] may need revision, as historical cases of posthuman convergences rely on their own historical episteme.

In the figure of these glass humans, early modernity has delivered an elaborate fantasy of an alternatively embodied human being who, while upholding cardinal points of humanism (the rational, knowing subject), dislodges its center in fundamental ways. The glass humans whose souls operate at higher speeds because of their glass embodiment and eschewal of flesh announce the failure of reason as it is assumed to operate, in the humanist paradigm, within and despite the body. Only in this sense can one conceive, perhaps, of the early modern glass human as a precursor to contemporary instances of the posthuman. But the early modern posthuman also offers a solution to humanism that is different and distinctively early modern in its conception of embodiment.

For those affected, the duplicity of melancholy was, indeed, a problem. That duplicity meant that one line of symptoms necessarily dragged along the other. Fear and sadness allowed for special intellectual skills, and those exceptional creative and thinking capabilities could always fall back into fear and sadness. In his work *De vita* (1489), a kind of self-help book devoted to scholars who wished to understand and better take care of their own melancholic inclinations,[126] Renaissance scholar and loyal member of the Medicean court Marsilio Ficino was very clear in connecting intellectual activity and melancholy: "Because the pursuit of knowledge is so difficult, it is necessary for the soul to remove itself from external things to internal things, as if moving from the circumference to the center. While one is looking at this center of man, . . . it is necessary to remain very still, to gather oneself at the center, away from the circumference. To be fixed at the center is very much like being at the center of earth itself, which resembles black bile."[127] The idea of mental activity as disengaging the body was philosophically latching on to Platonic dualism: "Then it is your opinion in general that a man of this kind is not concerned with the body, but keeps his attention directed as much as he can away from it and toward the soul?"—"Yes."[128] This was the maieutic exchange in Plato's *Phaedo*, where the philosopher is described as having to disconnect from the body to pursue much higher interests. Whereas in the pseudo-Aristotelian Book 30 of *Problems,* the exceptionality of (some) melancholics descended from the thermal properties of black bile—which could become (like iron) either extremely cold or extremely hot—in Ficino, intellectual absorption (desirable from the Neoplatonic perspective) possibly caused a split between mind and body and, through a sweeping analogical move, a movement toward the center of oneself—hence of earth, hence of black bile.[129] Further emphasizing this dynamic of separation, Ficino claimed: "Of all scholars, those devoted to the study of philosophy are most bothered by black bile, because their minds get separated from their bodies and from bodily things, they become preoccupied with incorporeal things, because their work is so much more difficult and the mind requires an even stronger will. To the extent that they join the mind to bodiless truth, they are forced to separate it from the body. Body for these people never returns except as half-dead and melancholic."[130] In Ficino, the formulation of the melancholic condition was rooted in the relation between the body and mind of the scholar in ways consistent with, though not identical to, the corporeal thinning of Cervantes's glass man. The idea of a close association of melancholy and intellectual activity, as we have seen, had been active since antiquity (at least since *Problems*) but was always inflected through different images. One of Cervantes's possible sources, the aforementioned Juan Huarte de San Juan, led the idea of impairment exercised by the body on the soul (which echoed with Vives's conception and with Platonism more at large) to extreme consequences when he ascribed to the bodily abundance of flesh the function of impairment and to emaciation (and some idea of the fasting body) the opposite power of genius: "Nothing offends the rational soul more than having to be in a body thick with bones, fat and flesh . . . and on the contrary, when the stomach is empty and thin, this greatly helps the genius, as we can see in the case of those who are hungry and needy."[131] In Ficino, the outcome of overcoming bodily impediments on the soul through intellectual effort was a forceful, and often dangerous, separation of mind and body, which like a wise doctor he had to warn against. In fact, Ficino's book offered a collection of recipes and concoctions for scholars to take care of their body, avoiding the detrimental effects of melancholy on health.

In the case of the glass melancholic, however, in lieu of separation (as of emaciation or physical deterioration), we witness a radical metamorphosis of the body that allows the mind to operate at higher speeds: a bodily transformation that grants an epistemic reorganization of mind/body dynamics. Humanism and its critique seem to converge here not in "the erasure of embodiment" predicated on some contemporary models of the posthuman—a disavowal of corporeal flesh.[132] A distinction and a divergence emerges here between the solutions offered by the humanist figure of the disembodied melancholic (as the scholar addressed by Ficino) and those presented by the glass melancholic. In the case of the latter, the embodiment in glass functions as a kind of solution to the perilous separation of body and soul caused by melancholy: a way to keep the potential dissociation at play by reinscribing it on different conditions of embodiment. A glass body does not disembody the human but reembodies it according to different, still very much physical, conditions. The glass body reasserts the body as a necessary dimension, even foregrounds it by making its features essential to the workings of the mind, while at the same time accomplishing and reversing the body/mind dualism on which early modern humanism rested.

However creative and brilliant, this solution was also necessarily painful; far from merely deserting the surrounding world, the glass human was constantly doomed to suffer from any contact with it. The body of the melancholic did not disappear but became a distinct site of convergence for multiple skills and performative negotiations between the glass human and society at large.

In the posthuman configuration of Vidriera's body, I read the map of a possible epistemology for early modern glass humans. The cognitive enhancement of the human achieved through a modification of the body grants the glass human a privileged place in the society that would otherwise marginalize him as a melancholic. While the body registers the pain of social exclusion through constant fear of breaking and a limited range of physical possibilities, the same body shields the glass human from external dangers by endowing it with a sacred untouchability. At one point in the story, Vidriera becomes furious at comments addressed against "a very fat clergyman" (*un religioso muy gordo*), and exclaims: "Let no one forget the words of the Holy Spirit: Touch not my anointed [*Nolite tangere christos meos*], and do my prophets no harm"[133]—a biblical formula found both in Psalms and in Chronicles.[134] The psalmic interdiction is one Vidriera would demand for himself—as a prophetic figure, in the socially assigned role of the satirist. Vidriera is asking not to be touched, lest he break, and his frailty thereby becomes a sign of consecration. Untouchable and sacred, the *licenciado*'s glass body functions as a protected zone, establishing a secure distance between itself and others (whether people or objects) and claiming a status of sacred exception.[135]

While the melancholic indicators of fear and sadness, and of a disturbed imagination, find an embodied reconfiguration in a glass body that, in its physical impediments, articulates the social significations of exclusion and estrangement from the social space, that body also becomes holy, sanctioning the exclusive zone of the disease and transforming the fear of breakability into a protective interdiction of touch. In the glass humans' syndrome, insofar as it is a form of melancholy, we should seek therefore not just the stigma of exclusion—of a painful interruption of the relations with the social space—but also the suggestion of a productive separation from the social body in the direction of an individual creation of meanings that is not exhausted by the possibilities of the "normal" and the "pathological." By precluding him from a wholesome social integration and by sanctioning his marginalization from the world, Vidriera's body of glass also reintegrates him into the social sphere as somebody whose mind works faster and better than others' because of the very same body that sets him apart.

Contrary to the posthumous superimposition of metaphoric meanings, acknowledging both the material and philosophical dimensions of embodiment of glass humans is the best available way to approximate the cultural tensions (with their provisory solutions) inscribed in this historical melancholic syndrome. It is also key to recovering the critical function of this figure that, by virtue of its embeddedness in the humanist episteme, poses a challenge to its stability.

The early modern melancholic imagination entertains a complex negotiation between the constant fear of a pathological separation of mind and body and the yearning for a productive solution, and it generates practical and figurative ways to project this negotiation onto the space of social exclusion, on the one hand, and of desirable exceptionality, on the other. The ultimate disappearance of the body is seen, however, within the melancholic discourse, as a fatal desertion—a risk to be avoided, a dangerous, vertiginous dream. The glass human takes shape at the intersection of the fear of and desire for disembodiment, articulating an alternative form of embodiment that arbitrates these tensions for the transient time of the glass human's delusion—the contingent time of Vidriera's *enfermedad*. For that interval, the glass human functions as a posthuman repository of power to exorcise the powerlessness of the melancholic subject as a by-product of humanist practices of social and epistemic regulation and, ultimately, to imagine an alternative way to resist while surviving, to dissent while vaticinating, to enhance the human condition without vacating it.

TWO

A Transparent Heart

GLASS HUMANS IN THE AGE OF ENLIGHTENMENT

In this chapter, I follow the exploits of glass humans as they were represented by some of the literary works that shaped the cultural moment of the Enlightenment. I do this by looking closely at their interaction with the scientific and medical discourse of the time in order to recover the role and function of glass humans within the epistemic project of European Enlightenment. What shall emerge is a figure that transgresses the boundaries of discourses to embody a critical function within the Enlightenment episteme.

If, in the first chapter, glass humans were entirely or partially made of glass, in this chapter they are endowed with a window on their heart. The glass human in the age of Enlightenment is a metonymical figure: what is fantasized as transparent is, with some variations, the site of the chest, or the heart—the figurative seat of feeling and, in traditional cardiocentrism, of thinking as well. The Enlightenment thereby performs a reduction of the figure of the transparent human to its operative centers. As a metonymic focal point, the transparent heart, through its selective confinement, emphasizes, on the one hand, the figurative function of the trope and, on the other, its epistemic charge. The localization of the glass pane on the site of the cardiac muscles shifts the epistemic focus of these figures from the problem of fragility and social integration onto the problem of introspection. The fictional literary figures I will dwell on in what follows are well-crafted tropes designed to rethink conditions of embodiment in relation to the stakes posed by the enterprise of knowledge pursued at multiple places in the European eighteenth century.

The traditional historiographical understanding of Enlightenment culture from the standpoint of its epistemic ambitions is that of a "visual primacy" that organizes multiple kinds of scientific, as well as poetic, enterprises attempted by the authors working under its framework.[1] The glass humans we will encounter in this chapter are figures who both display and challenge these visual ambitions, embodying the tensions between the primacy of the visual and the demands of what is nonvisual or beyond the visual. Thus, they fulfill a similar posthuman function as that of their early modern antecedents, pursuing an epistemological questioning of humanism around the "anthropological junction," as Albrecht Koschorke has called it, of European Enlightenment.[2]

The first text I will discuss, Christian Heinrich Spieß's novella *Die Geschichte von Jakob W***r* (The story of Jacob W***r) (1796), ties the figure of the glass melancholics from the previous chapter to the other glass-hearted humans of the present one. The novella stages the figurative imbrication of the glass melancholy syndrome and the trope of the transparent heart in connection with the revamping of physiognomy, a (pseudo-)science that the Swiss Pietist minister Johann Caspar Lavater had systematized in his *Physiognomische Fragmente, zur Beförderung der Menschenkenntniss und Menschenliebe* (1775–78).[3] Close to the problems of physiognomy, and yet with a different epistemic focus, are the two texts at the center of the chapter: Jean-Jacques Rousseau's *The Confessions* (written between 1765 and 1769 and published posthumously in 1782) and Laurence Sterne's *The Life and Opinions of Tristram Shandy, Gentleman* (published between 1759 and 1767). An autobiography and a fictional autobiography, respectively, the two works act as foils for one another—the earnestness of Rousseau's authorial confessions serving as a counterpoint to the facetiousness that suffuses the fictional narration of Tristram Shandy's life. Both works engage, in different ways, the power of literary discourse in giving account of someone's life, and both interrogate the means of the autobiographical (and more or less fictional) form to achieve introspection, but with rather opposite results. Where Rousseau strives to prove his reliability and the success of the act of self-revelation, Sterne stages the constant defeat of truthfulness, honesty, and self-understanding in literary writing and life alike.

At crucial points of this shared yet antithetical exploration of literature's mastering of introspection, both Rousseau and Sterne draw on the image of a transparent heart as the figurative hypothesis that can sanction literature's claim to capturing the true nature of its subject and, more to the point, penetrating someone's inner life and accounting for it.[4] These literary images, far from simply being rhetorical tropes, intersect with models and representations of the human body and the human being in multiple, interwoven discourses. Furthermore, these images not only interrogate the scope of literary introspection but pose the larger anthropological question of how to achieve knowledge of a human being's interiority on a physiological, as well as moral, plane. I thereby bring the intersection of literature and medicine to bear on the understanding and interpretation of these figures, selecting images and objects elaborated in the medical discourse that sustain the plausibility and efficacy of the literary tropes.

The critique of humanist mainstays performed by the figures of this chapter in the form of imagined alternative notions and formats of humanity charts a critical, posthumanist-inflected undoing of the subject of the Enlightenment episteme, thereby sharing a critical function with the other transparent humans whose stories are told in this book, which all implement fictional strategies to articulate, represent, or reimagine the condition of being human.

*The Man with a Window on His Heart: Jakob W***r*

In 1796, bestselling German author Christian Heinrich Spieß, son of a pastor from the eastern region of Germany, published a collection of case studies of mad people (*Biographien der Wahnsinnigen*), a catalog of entertaining stories with anecdotical, as well as moralistic, underpinnings.[5] The story I am concerned with is that of "Jakob W***r," a melancholic glass man who allegedly lived almost two centuries after Cervantes's *licenciado Vidriera*.

In the Alpine region of Tyrol, a twenty-four-year-old man named Jakob has inherited money that he has invested with great profit, thanks to his intelligence, various talents, and social skills. The young man falls in love with his housemaid, Marie, and falls prey to a typical form of what Robert Burton had described, in his *Anatomy of Melancholy* from 1621, as love melancholy. Since Jakob hails from an upscale and traditionalist family, he feels he would raise a scandal if he were to marry the maid, marriage being one of the main tools of social control and class order. After having confided his feelings only to one trusted friend, Jakob attends a village marriage where, to his utter surprise, all his relatives seem to be already informed about his secret love and directly reproach him for his "shameful love" (*schandvolle Liebe*).[6] Jakob is less shocked by the negative and hostile reactions of his relatives, which he, in fact, had anticipated, than by their sheer knowledge of his intimate thoughts. Terrified by this turn of events, he leaves the premises and disappears for a week. Believed to have perished in the cold winter, he is found one week later by his servants in a hay shed on the mountains, his hands tightly closed on his chest.[7] Over the following days, Jakob never lets his hands off his chest except for one time, in the presence of a priest and, a second time, when his beloved Marie visits him. On that occasion, Jakob asks the woman to just read his love: "Just look! Just read!" From her only, he adds, he need not hide his heart.[8] As the woman seems confused, he goes on to explain that a great transformation has occurred in his body: "My breast has become a lantern!" Now everybody can see his thoughts as through glass: "Isn't it here and here out of glass? Do you see my heart and all my thoughts inside?"[9]

The doctor who takes care of him, informed of Jakob's belief, complies with the fantasy (unlike one doctor seen in the previous chapter) and promises him a therapy through a special compress to be placed on his chest to allow flesh to grow back. The skilled physician who hears this story offers then a complex diagnosis, involving love melancholy, effects of wine, exposure to the cold, and the structure of imagination [*Einbildungskraft*], which he compares to a moving wheel, a magic lantern that can get sometimes stuck on one image.[10]

The therapy devised by this wise doctor, however, will eventually fail, leaving Jakob to be a man with a glass chest until his death. Jakob will never get out of his delusion: after several years, however, he has become a renowned card player with an unbroken winning streak and has regained control of his business with skillful management (from inside the house). Never does his managerial judgment fail, and he even starts performing prophetic tasks as it regards the weather. Last but not least, when he talks to his employees and servants, he always knows whether they are telling the truth. Not unlike Vidriera and the other glass humans of early modernity, Jakob seems to turn his condition into an asset, confirming his melancholic allegiance to a tradition of extraordinary humans. A *perittós*, as in the pseudo-Aristotelian text of *Problems* analyzed in the previous chapter, and therefore endowed with extraordinary cognitive and even prophetic capacities, Jakob reembodies the early modern tensions at the core of the melancholic complex while intercepting the emergence of new discourses. As powerful as his unique skills are, his folly (*Wahnsinn*) will never leave him; on his death bed, he will pray that no one open his coffin, so that nobody might read his heart.[11]

But how could Jakob get such a heart of glass in the first place? Trying to reconstruct the chain of events that have led him to this transformation, Jakob had told the doctor how, right after fleeing from the disgraceful marriage event, the ghost of his own mother appeared to him in the shape of a white lady, a "weiße Frau," a typical figure of German gothic folklore and herald of death, who shamed him for having a heart of glass and then exhorted him to hide: "Shame on you child, are you trying to disgrace me? You've got a breast of glass. . . . Hide!"[12]

The plot of shame that psychoanalysis has trained us to read in between the lines of such a tale is obviously still far from Jakob's world of meanings. If affect theory hinging on psychoanalysis ascribes shame to a universal order of the gaze,[13] whether socially externalized or individually introjected, cultural historical readings of this complex affect insist on its social semiotics, where reference values are constantly changing.[14] In Jakob's story, shame—*Scham* and its external correlate *Schande*, translatable as "disgrace," the social effect of shameful behavior—is sanctioned by societal rules of appropriateness, and Spieß's *Biographies* attests generally to the extent of the culture of shame that ran deep in the Protestant society of Germany at the end of the eighteenth century—that culture that Spieß, who had grown up in the Electorate of Saxony, knew intimately—as well as in the presumed setting of this story, the Habsburg, Catholic province of Tyrol, with a minority of Protestants. In fact, exposure to public reprobation is the thread of almost all the unfortunate protagonists of Spieß's *Biographies*.[15]

It is ultimately shame that provokes Jakob's bodily metamorphosis. Shame is culturally regulated, yet its experience is bodily inscribed, pushing the subject of shame to question their bodily presence and, with it, their existence. Shame opens up an undoing of the subject that nobody has captured better than Emmanuel Levinas, when, writing on modes of "escape" for an untenable Western philosophical subject, he wrote that "what shame discovers is the Being who uncovers itself."[16] While society prescribes in always changing circumstances and norms the opportunity of shame, shame is an affect experienced strongly through the body: a corporeality that seems to attach the various possible cultural codifications of shame to an ontological ground in which the uncomfortable contact with one's weakness and helplessness in front of one's own creaturely mortality is culturally experienced.[17]

But the affective dimension opened up by shame, which, following such analyses, may be capable of mobilizing the ontology of the subject, is nonetheless accessible to historical cultural analysis as discursive, produced within several concurrent discourses. One of these is decisively to the cultural junction of Jakob's life and is that of physiognomy.

Jakob's tale has been read as a reaction to the anxieties unleashed by the science of physiognomy, which was peaking at the end of the eighteenth century with the reception of Lavater's work. Alexander Košenina has shown how around 1800, the image of a window on the heart functions as the negative correlative to disclosure and exposure of thoughts and emotions promised by physiognomy.[18] The rest of Spieß's *Biographies* does indeed confirm a broader use, in the speech of his characters, of the figure of a transparent heart as metaphor for exposure of inner thoughts and intentions, often however in positive terms as figure of truthfulness.[19]

If physiognomy provided technical tools that seemed to threaten the protection of secrecy (Goethe called it after all a "spying business"),[20] it also aligned with a diffuse semantics of an *eloquentia cordis*,[21] which spanned religious and political practices, with the anthropological aspirations of an age, and with some of its scientific ambitions. In fact, that window on the heart was already looming large when Jakob W***r got one of his own on his chest.

The Transparent Heart Between Literature and Medicine

Why did both Rousseau and Sterne, in their competing yet complementary forms of autobiography in the age of Enlightenment, use the image of the transparent heart? Is it a mere coincidence that they both resorted to it, or is it rather a deeply motivated alignment? What happened around 1760 to justify the relevance of the image of the transparent heart as a crucial figure for the interrogation of the reach and power of literature in disclosing its characters and their inner lives? Trying to answer these questions, I will show how a precise development in scientific discourse, involving medicine, physiology, and anatomy, sustains the relevance of this image in the metaphorical language of these authors.

Both Jean-Jacques Rousseau and Laurence Sterne, who had written their texts just a few years before Spieß, are deeply concerned with the same question that concerns Jakob: that of visibility and access to one's inner life.

Their concern, however, is not pathologized but rather put to the service of a hermeneutic of introspection that, instead of relying on the exceptional possibilities entailed in a still pathological setup, build on the epistemic ambitions of contemporaneous medical discourse. By integrating literary and medical discourses, I want to emphasize the dynamic relationship between scientific models of the body and literary poetics in the construction of the imagination of transparent human figures.

While this is a methodological foothold for literary scholarship interested in the heteronomy of literature (that is, literature's permanent exposure to, and interaction with, other discourses that fundamentally shape its motives and its aims),[22] it is also an inherently historical stance. During the Enlightenment, the interaction between literature and medicine is distinctly manifest as overlapping topics of interest and intellectual concerns, to the point that the place of medicine in Enlightenment culture at large could be seen as that of a "master discourse."[23] More than merely influencing literature (in ways easily proven by the familiarity of the literary authors under examination with scientific and medical texts), Enlightenment medicine shares with literature basic epistemic ambitions.[24]

In the literary image of humans with a transparent heart, a composite scientific picture of the body is crucially mobilized, comprising anatomical, physiological, and medical discourses, among others. The image of the transparent heart evokes, in fact, at a primary visual level the picture of what, by the end of the seventeenth century, had become the prevailing mode of visualizing the body: the mechanist model of the human machine. The mechanist wave in medicine, with its different roots in England and Italy, and its different paths—from Descartes's speculations to Malpighi's dissections—flooded Europe for almost a century. In this chapter, I will therefore spend some time on Mechanism's impact on the cultural imagination that also produced the figure of the transparent heart.[25]

By the time Rousseau and Sterne were writing, however, profound changes had occurred in the science of the body, many of which had explicitly or implicitly undermined the mechanist perspective.[26] At the time of Sterne and Rousseau, a process of uncovering the simplifications and limitations of the mechanistic explanations of bodily functions had begun and was fully running, heralded by what is traditionally seen as the school of thought opposed to Mechanism—namely Vitalism, a highly diversified field of knowledge that involved anatomy, physiology, and medicine.[27] As the themes of Vitalism were becoming prominent, the picture of the body that was starting to emerge from this new branch of knowledge was far from mechanistic: rather than being understood as a clockwork beneath the skin, the body started to lose its transparency.

Why, then, hold on to the picture of the transparent heart? What function did this image have, at that point? This tension between old and new models attests, on the one hand, to the visual, scopic yearning of Enlightenment authors—one of the touchstones of Enlightenment historiography, as I have mentioned—but, on the other hand, it undermines it by letting it enact and demonstrate its crucial epistemic shortcomings. Sterne and Rousseau had thoroughly assimilated Mechanism: without the mechanical models of the body reviewed in the following pages, the image of an aspirational narrator looking at his characters by metaphorically opening their chest (Sterne), or that of an autobiographer exposing the transparency of his own heart to his readers (Rousseau), couldn't have arisen. On the threshold of the second half of the eighteenth century, however, both authors were exposed, albeit in different ways, to the novelty of a different model of the body—a model that was not fully established yet and therefore not completely available. The new Vitalist notion of sensibility introduced, as we will see, the nonvisual quality of the vital processes taking place in the body, doing away with the visual machinery constituting the mechanist body.

As writers and novelists, Sterne and Rousseau were dealing with this transition at a different pace and toward different ends. On the one hand, Sterne was strongly critical of the reductionist representation of the mechanical body and tried to show its limitations by making fun of mechanist medical discourse (among many other vicious scientific traps of his time). Rousseau, on the other hand, who took his confessional writing very seriously, was greatly affected by the new Vitalist wave of "sensibility" and tried to picture himself as the sensitive man *par excellence*. To do that, he drew on something at odds with the science of sensibility as a product of Vitalism—namely, a mechanistic picture of the body: the image of the transparent heart. In so doing, he showed how the new discourse of the body was still in search of a viable representation; and reusing what was already available, trying to bend it toward the new Vitalist meanings that were now being attached to the human body, he ended up with a figure that entailed its own critique. Foils of each other, these two authors, by returning to variations of the transparent human, interpreted and problematized the cultural and scientific tensions of their historical moment.

The Mechanics of a Transparent Heart: Sterne's Dioptrical Beehive

Chapter 23 of the first volume of *The Life and Opinions of Tristram Shandy, Gentleman* contains a complex, ironic digression on the transparency of the heart. In this instance, the narrator's departure from the main story hinges on a double fantasy of the human body: in the first part, a body endowed with a window, or an opening into a "dioptrical bee-hive," and in the second part, a nonhuman (Mercurian) body entirely vitrified. Both fantasies are introduced as explicitly fictional and are meant to prove by means of their absurdity (they are, in Sterne's words, "nonsensical") that the very notion of a transparent heart is not only unviable but even ridiculous:

> I have a strong propensity in me to begin this chapter very nonsensically, and I will not balk my fancy.—Accordingly I set off thus.
>
> If the fixture of Momus's glass, in the human breast, according to the proposed emendation of that arch-critick, had taken place,—first, This foolish consequence would certainly have followed,—That the very wisest and the very gravest of us all, in one coin or other, must have paid window-money every day of our lives.
>
> And, secondly, That had the said glass been there set up, nothing more would have been wanting, in order to have taken a man's character, but to have taken a chair and gone softly, as you would to a dioptrical bee-hive, and look'd in,—view'd the soul stark naked;—observ'd all her motions,—her machinations;—traced all her maggots from their first engendering to their crawling forth;—watched her loose in her frisks, her gambols, her capricios; and after some notice of her more solemn deportment, consequent upon such frisks, &c.———then taken your pen and ink and set down nothing but what you had seen, and could have sworn to:—But this is an advantage not to be had by the biographer in this planet,—in the planet Mercury (belike) it may be so, if not better still for him;——for there the intense heat of the country, which is proved by computators, from its vicinity to the sun, to be more than equal to that of red hot iron,—must, I think, long ago have vitrified the bodies of the inhabitants (as the efficient cause) to suit them for the climate (which is the final cause); so that, betwixt them both, all the tenements of their souls, from top to bottom, may be nothing else, for aught the soundest philosophy can shew to the contrary, but one fine transparent body of clear glass (bating the umbilical knot);—so, that till the inhabitants grow old and tolerably wrinkled, whereby the rays of light, in passing through them, become so monstrously refracted,——or return reflected from their surfaces in such transverse lines to the eye, that a man cannot be seen thro';—his soul might as well, unless, for more ceremony,—or the

trifling advantage which the umbilical point gave her,——might, upon all other accounts, I say, as well play the fool out o'doors as in her own house.

But this, as I said above, is not the case of the inhabitants of this earth;—our minds shine not through the body, but are wrapt up here in a dark covering of uncrystalized flesh and blood; so that if we would come to the specifick characters of them, we must go some other way to work.

Many, in good truth, are the ways which human wit has been forced to take to do this thing with exactness.[28]

This passage builds a complex digression to motivate Sterne's strategic approach to the description of characters: not up front but, as it were, from the side—namely, from external elements of behavior or what in the novel is called "hobby-horses."[29] According to the narrator, a more direct description of the characters, starting, say, from their inner motions and machinations, would be impossible for the simple reason that the "heart" of humans is "wrapt up in a dark covering of uncrystallized flesh and blood."[30] By mentioning explicitly the tale of Momus, the narrator locates his digressive hypothesis *per absurdum* in the dimension of fable or, more exactly, that of myth. Momus had reproached Zeus for failing to build a window on the human chest—that is, to provide humans with transparent minds or hearts (according to alternative versions of the myth).[31]

To prolong the sense of this impossibility, against which Sterne's narrator propounds his own narrative strategy, a second digression follows, again *per absurdum*. What would it be like to not just be provided with a glass window on the chest but to be entirely made of glass? The narrator develops a fantasy about the inhabitants of Mercury as creatures vitrified by the scorching temperatures of their planet. The imagined "transparent body of clear glass" of the Mercurians would allow the soul to be fully visible on the outside, at least until the wrinkles on the aging glass bodies might start to refract and reflect the light to the point of blinding the onlooker, invalidating the very transparency of their bodies. This curious fantasy of a blinding, reflecting body of glass functions to undermine the idealization fostered by the complementary fantasy of an entirely visible inside of the human: the narrator is warning that the transparency of this inside (here thought of as the soul) may well be an advantage for the onlooker only as long as it is not taken too seriously, in which case it may just merge into its opposite: blindness.[32]

Sterne's dioptrical beehive is, however, a figure more complex than that of a transparent heart or of a glass window on the chest. On the one hand, the qualifier "dioptrical" entails the magnifying action of a lens—a microscopy effect; on the other hand, the "bee-hive" conjures a peculiar structure for this mechanism: a living nest where the life of "maggots" would become visible thanks to the magnifying effect. Kate Tunstall has defined this particular image as "a deliberate textual staging of the vertiginous effects of microscopy."[33] Microscopy and Mechanism went, in fact, hand in hand for at least some crucial part of the history of Mechanism.[34] Robert Hooke's 1665 *Micrographia*, an atlas of illustrations from microscopic observations, notoriously had a groundbreaking impact on the reading public at large, and it certainly had a fascinating effect on Sterne; I believe it might have even been a direct source of inspiration for Sterne in the passage we saw—in particular, a passage from the *Micrographia* where a special kind of marine "bee-hive," with its peculiar generation of maggots and gnats (i.e., not bees, but mosquitos) is described and illustrated (fig. 7).[35]

As Bertoloni Meli has argued, Hooke had observed these insects (the

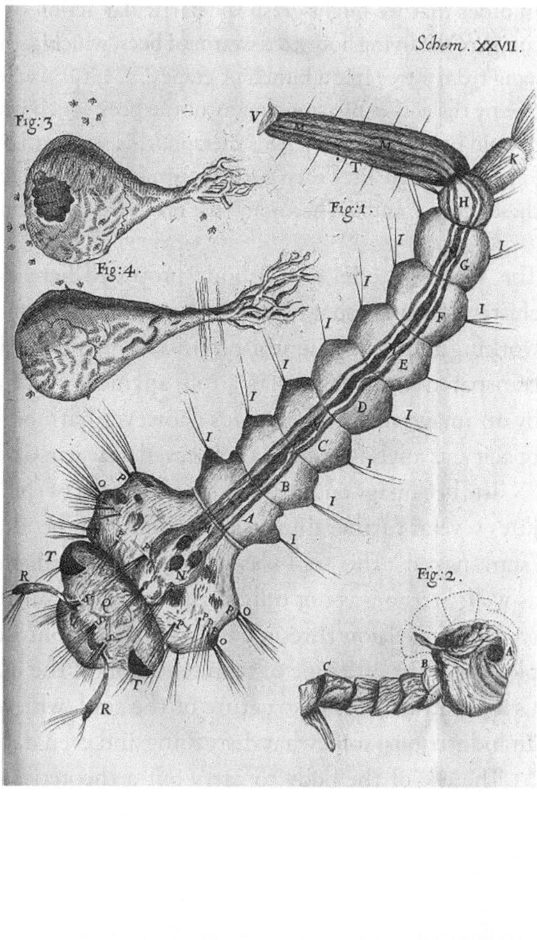

FIGURE 7. Robert Hooke, "Apiarium Marinum," in *Micrographia: or, Some Physiological Descriptions of Minute Bodies Made by Magnifying Glasses* (London, 1665), table 27, p. 186. Smithsonian Libraries and Archives.

gnats) through their transparent bodies, very much as William Harvey had done with the hearts of Thames shrimp when he was researching the circulation of blood, leading to his great "discovery" of the heart pump.[36] The imagery of maggots, however, strongly evokes images of rotting flesh, and the action of "going softly" to open this mechanism introduces the idea of a possible danger coming from these insects, if one were to disturb them, which adds a layer of hazard to the whole eagerness to find out about the inner life of humans.

There is a further way the image of the beehive challenges the aspirations to a mechanic transparency of the heart. Once mechanistic models had become unsatisfactory, different models for the body were conceived to account for the specificity of the living body as opposed to others. One of the leading voices of Vitalism in the eighteenth century, the French doctor Théophile de Bordeu, wrote in his *Recherches Anatomiques sur la Position des Glandes et sur leur Action* (1751):

> In order that we might grasp the particular action of each of its parts better, we compare the living body to a swarm of bees, which gather together in a cluster, and hang from a tree like a bunch of grapes . . . ; it is a whole glued to the branch of a tree by the action of a good many of the bees which need to act in unison in order to hold on. . . . Applying [this metaphor] is straightforward: the bodily organs are joined together; they each have their own area and action. The relation between these actions, and the harmony that results, is what constitutes well-being.[37]

The Vitalist model of the body proposed here evokes a swarm of bees, a cluster of beings that exist and function in harmony similar to the organs working and acting "in unison" in a human body. In this model of the body, transparency does not play a role anymore and has actually been replaced by an image that evokes chaos (however harmonious) and a dense kind of opacity, as anybody who has observed a swarm of bees knows.

The bee image encapsulated in the image of the "dioptrical bee-hive" conjures several further dimensions, adding to the idea of a soul to be observed "stark-naked": this soul was not just a mechanism but a peculiar organism as well. Sterne was not only suggesting the absurdity of the hypothesis of a transparent vision through a glass window onto the human heart; by employing a Vitalist trope of representation of the human body, he was giving us a glimpse into the structure of the soul, which, if seen, would look like an industrious, somewhat disturbing and even dangerous, swarm of bees.[38]

The use of the body to carry out a theoretical and rhetorical statement about the narrator's knowledge of his characters is neither surprising nor the only one in a novel, such as *Tristram Shandy*, that is constantly concerned with the body and its diseases. Judith Hawley has written about how Sterne's proximity to the themes of Mechanism might explain the prominent place mechanist representations hold in the novel; in fact, as she argues, mechanics seems to govern the world of *Tristram Shandy* at the very level of its characters' behaviors.[39] At the same time, however, as one might expect from a consistently ironic work, the use of mechanistic theories is exposed to continuous parody and comical effects, intended to expose contradictions rather than argue for an antagonistic stance, such as pitting Vitalism against Mechanism.[40]

Sterne's knowledge of contemporary medicine has been extensively assessed, and traces of the medical discourse of his time are copious throughout the novel.[41] Not only medicine is a central concern and theme in the Shandy household, but Sterne is highly familiar with late seventeenth- and early eighteenth-century iatromechanics.[42]

At the end of the "dioptrical bee-hive" passage, the narrator concludes that he will describe the character of uncle Toby "by no mechanical help whatever" and that at the same time he will also abstain from the empirical clinical practice of examining bodily "repletions" or "discharges."[43] In a single move, Sterne manages to dismiss the medical methods of Mechanism and Empiricism as diagnostic tools to understand and represent both the human body and character. In the novel, the body is indeed exposed as a machine, but it is one that works consistently badly, where disease is constantly threatening life.[44] In a novel where "everything tends eventually toward a bodily and sexual inference,"[45] the reference to the "dark covering" of "flesh and blood" is therefore not just a description of the body as such; it fundamentally defines the relation between knowledge and its object, specifically between abstract narrators and characters who are entangled in bodily existence and trapped in their corporeal predicaments. Narrators may wish to look into their characters as if they were perfect mechanisms, or dioptrical beehives, but this turns out to be a ridiculous aspiration, the absurd consequences of which are manifested in the passage in question.

In this sense, Sterne's endeavor is the opposite of that of the physicians who were describing a man-machine in those years, whom Sterne both cites and mocks in his novel, ridiculing their "delusion of medical prometheanism."[46] Sterne's use of the popular medical Mechanism of his time fits the larger project of his work as stated in the famous letter he wrote to his publisher, Robert Dodsley, in 1759, in which he made very clear that he intended his book as a sweeping satire of contemporary scientific discourse: "The Plan, as you may perceive, is a most extensive one,—taking in, not only, the Weak part of the Sciences, in which the true point of Ridicule lies—but every Thing else, which I find Laugh-at-able in my way."[47] In the constellation of characters in the novel, it is especially Walter Shandy who embodies a paradigmatically blind faith in science, with ridiculous outcomes that end up sabotaging any alleged scientific credo.[48]

Sterne's particular take on embodiment was at odds with the dreams of introspection that saturated Enlightenment literature in the form of its confessional thrust, as I will show with the example of Rousseau. In a brief passage from Sterne's chapter 17, located in the fourth volume of the novel, not only does the narrator parody Walter's inability to apply reason to practice but categorically disavows the possibility and value of introspection: "We live amongst riddles and mysteries, the most obvious things, which come in our way, have dark sides, which the quickest sight cannot penetrate into; and even the clearest and most exalted understandings amongst us, find ourselves puzzled and at a loss in almost every cranny of nature's works."[49] Though rhetorically framed in a highly ironic utterance, this passage, in fact, settles the question of transparent knowledge once and for all: there is no such thing as "penetrating" with sight the meaning of a human being, let alone of nature's works; everything is muddled, riddled, opaque. With *Tristram Shandy*, Sterne is mocking the larger epistemic program of the Enlightenment, which spans its scientific (especially medical) and anthropological goals.

Human Machines: Transparent Models of the Body

In the analyzed passage on the "dioptrical bee-hive," Sterne borrows from the imagery of Mechanism in order to ridicule it. Though the use of machine-models had long been a means for explaining physiological processes,[50] the idea of a thoroughly mechanical body had only properly taken off in the second half of the seventeenth century, which became "the golden age of mechanistic anatomy."[51] Mechanist models of the human body multiplied, proposed by anatomists and doctors, and further perused by philosophers. René Descartes, in his *De homine* (1662), one part of his planned but unfinished opus *Le Monde*,[52] set out to describe humans as machines created by God:

> I assume their body to be but a statue, an earthen machine formed intentionally by God to be as much as possible like us. . . .

> We see clocks, artificial fountains, mills, and similar machines which, though made entirely by man, lack not the power to move, of themselves, in various ways. And I think you will agree that the present machine could have even more sorts of movements than I have imagined and more ingenuity than I have assigned, for our supposition is that it was created by God.[53]

From the hydraulics of circulation and the mechanics of digestion to the particular movements of "animal spirits" through the brain, Descartes subsumed the entirety of bodily operations under mechanics.[54] Indeed, he was surrounded by mechanistic theories of the body, of its physiology and its pathology.[55]

In the lively milieu of Italian medicine, with its dual headquarters in Bologna and Pisa, prominent anatomists and microscopists discussed how to better represent the human machine. Within the context of a debate over the relative virtues of mechanistic and empirical medicine, Italian doctor Marcello Malpighi (1628–94) described the human body as a system of machines "composed of ropes, wires, planks, levers, canvas, flowing fluids, tanks, channels, felts, valves, and similar machines. Man, examining these parts through anatomy, philosophy and mechanics, has learned their structure and their uses, and proceeding *a priori*, has formed models of them, through which he can visualize the causes of certain effects, and explain them *a priori*, and thanks to these, helped by discursive faculty, understanding the way nature operates, he founds physiology, pathology and later the art of medicine."[56] Malpighi provided very concrete examples of such corporeal machines—for example, the *camera obscura*: a mechanic model of human sight, the physiological and pathological functioning of which could be demonstrated a priori by observing the optical model itself. Moreover, the machine-model "reveals *a priori* the way of nature" in the case of blood circulation, the skeletal apparatus, respiration, digestion, and so on.[57] The body, for Malpighi, was subject to the laws of nature, which were the laws of mechanics; thus the doctor, "through the laws of movement and the variety of figures, deducts *a priori* the structure of these machines and the composition of those fluids."[58]

Drafted by doctors, mechanical models of the body were produced pervasively in the course of the seventeenth century. For example, Giovanni Alfonso Borelli (1608–79), a colleague of Malpighi's in Pisa and the leading figure of Italian Iatromechanism, described the human body in mechanical sequences that again involved levers, weights, and forces.[59] And one of Malpighi's students, Giorgio Baglivi (1668–1707), laid out even more graphically the mechanical structure of the human body: "For, if anyone carefully examines the structure of it, he will certainly find in the jaws and teeth a pair of pincers, in the stomach a container, in the veins, arteries, and other ducts water channels, in the heart a piston, in the intestines a sieve or filter, in the lungs bellows, in the muscles the power of the lever, in the corner of the eye a pulley, and so on."[60]

The description of such human machinery inevitably recalls a more famous one, which has come to epitomize the excessive machine-inspired fantasies of Mechanism: that of Julien Offray de La Mettrie (1709–51), who, just a few years before Sterne's novel, had written that man was a watch constructed with skill and ingenuity:

> I am not mistaken; the human body is a clock [le corps humain est une horloge] but so huge and cleverly constructed that if the cog which tells the seconds happens to stop, the one which tells the minutes goes on turning, in the same way as the cog for the quarters continues to move, and so do the others, when the first ones are rusty or out of order for some reason and stop working.[61]

This theoretical metaphor was taken quite literally by La Mettrie, for whom a good doctor had to treat the body as a machine:

> If we compare two doctors, the best and most trustworthy is always, in my opinion, the one who knows the most about the physics or the mechanics of the human body and who, forgetting the soul and all the worries which this figment of the imagination causes in fools and ignoramuses, concentrates solely on pure naturalism.[62]

L'Homme machine (1748) is often regarded as a manifesto of Mechanism, albeit La Mettrie was actually proposing such a model as a statement in materialistic ontology, against any metaphysical allegiance.[63] The conclusion of this idiosyncratic and ambiguous Mechanistic manifesto was, in fact, an appeal to the unifying matter of the universe: "Let us then conclude boldly that man is a machine [l'Homme est une Machine] and that there is in the whole universe only one diversely modified substance."[64] The ultimate goal of La Mettrie's mechanic philosophy was the integration of Mechanism in a materialist theory of the human and ultimately a different view of the human body than that pursued by Mechanism alone.[65]

The pervasiveness of mechanistic models of the body was so vast that it is not surprising that beyond these speculative or didactic descriptions, actual objects were built, which could display the mechanical workings of the human body. One such object was designed by German doctor Salomon Reisel (1625–1701) and was known to Malpighi as well (figs. 8 and 9).[66]

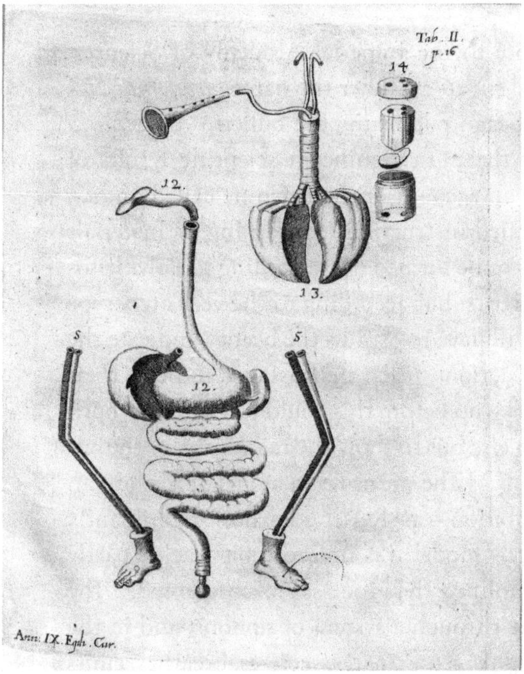

FIGURE 8. Salomon Reisel, *Statua humana circulatoria* (1680). Leopoldina, Nationale Akademie der Wissenschaften, Halle (Cb 4:400).

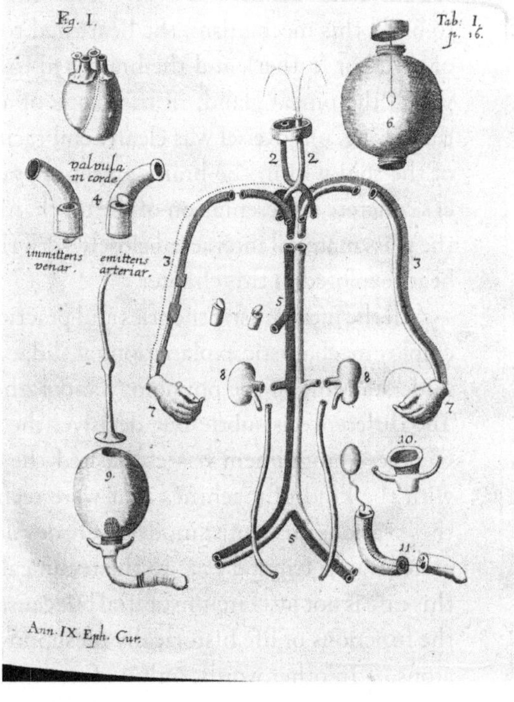

FIGURE 9. Salomon Reisel, *Statua humana circulatoria* (1680). Leopoldina, Nationale Akademie der Wissenschaften, Halle (Cb 4:400).

In 1674, Salomon Reisel wrote to the Imperial Academy of Sciences, which had recently been founded (in 1652) under the name *Academia Naturae Curiosorum* and, in 1670, began publishing the bulletin *Miscellanea curiosa*, which released some of the most significant scientific results of the time.[67] In 1678, Reisel wrote an account of the machine called *Statua*, along with illustrations and detailed instructions concerning its materials and structure: "Thus our statue is to be formed not according to Polykleitos canon, which is a norm for the artists; but according to the very archetype of the human body, of which it will have to express the beauty, indicate the tricks, list the parts, describe the actions, teach the passions, indicate their appearance: the ocean may well dry up before this could be done appropriately. But our aim here is nothing else but to show the mechanics of the circulation of blood, chyle, and serum."[68] The statue was a model of the human body, without any aesthetic ambition—as Reisel somehow redundantly points out. The appearance of this model was indeed a bit wacky, partly because Reisel had drastically simplified the processes it represented.[69] The Statua, in any case, was operating through a system of siphons, and in the words of his maker, it "lived": "it pulsates, urinates, secretes, breathes, emits sounds, talks, bends, and lives."[70] According to Reisel's instructions on how to build this mechanism, the heart had to be built of glass (or alternatively of metal or leather), and the brain had to be a glass sphere, at the center of which the pineal gland, in the shape of a crystal, was to hang from a silk thread (fig. 9):[71] Reisel was clearly embracing Descartes's physical placement of the soul within the brain, and more precisely in the pineal gland. Reisel's concrete representation of the mechanics of the heart and brain through the glass material interacted closely with the iconography of the transparent heart explored in this chapter.[72]

All the mechanistic theories and practices I've presented so far do not just employ mechanistic explanations as didactic models but represent an actual understanding of the physiology and pathology of the body as a machine.[73] The difference is subtle but decisive: the body *is*, ultimately, a machine.[74] Georges Canguilhem has connected the historical success of Mechanism with the kind of machines that were technologically available at the time these various mechanist models were developed. For Canguilhem, even Descartes was "a tributary . . . to the technical forms of his age."[75] The claim in this case is not just circumstantial but causal: "the mechanical explanation of the functions of life historically presupposes . . . the construction of automatons."[76] In other words, only a motor machine developed through technological means could lead to the development of the theoretical model of the machine-organism. The fate of Reisel's *Statua circulatoria* brings, indeed, this causal association full circle. After its construction, the machine was located in the *Wasserspiele* (fountains) of the Schlosspark in Stuttgart as the figure of Cupid.[77] Through this relocation, the *Statua* joined the world of automata, which reached a peak in production in the sixteenth and seventeenth century, in particular with the type of android automaton seen in the models built by Jacques Vaucanson (1709–82).[78]

Canguilhem's suggestion that representational models depend on technological practices helps explain how literary tropes (such as the transparent heart) are also bound to the existence of representational as well as technical objects (the mechanical model, the actual model of the "Statua"). Reisel's mechanic model posits the correlation, at the core of this chapter, between the mechanic model of the body and the transparent body. The mechanic body not only has to work as a machine, but its machinations have to be visible—hence, the use of glass to display the operations both of the heart and of the brain. The mechanistic conception of the human body implies a strong emphasis on the mechanism itself, aesthetically stylized in its display. If this is not really the case with the android automata, which are designed to delude the audience about their humanity, it is certainly the case of Reisel's *Statua*, which intends to reduce the mechanics of the body to that of a machine, therefore displaying emphatically a system of pistons, syphons, and levers. The body is made understandable and knowable by being made transparent: Reisel's *Statua* epitomizes the high hopes and theoretical reach of Mechanism. The transparent mechanism of the body is the construct that subtends Sterne's image, and its frank optimism in the possibility of controlling that body by representing its inner workings underpins Sterne's satirical fervor.

Against the Machine: Vitalist Models of the Body

Iatromechanism was not the only path down which medicine traveled in the crucial years of the eighteenth century when the image of a human with a transparent heart made its appearance in the narratives of Sterne we have just seen and Rousseau, as we will see. The historiography of science has traditionally posited a rivalry between Mechanism and its counterparts, namely what has come to be known as the "sciences of life."[79] This tension has also been schematized within the history of science as an opposition between the "champions of the primacy of nature" governing humans together with all that exists and "the upholders of the primacy of man."[80] The genealogy of this simplifying division has been equally debated: while some claimed a distinct genesis for the sciences of life, others emphasized their hybrid origin, pointing, for example, to the influence of Leibniz and Newton themselves in shaping anti-Mechanist, Vitalist theories of the body informed by the notion of immanent corporeal forces.[81] Eschewing schematic historiographic constructs and returning to the historical moment at stake, what emerges is a far more nuanced tension between different epistemologies that are not only "nonoppositional" but thoroughly interlaced.[82]

What all opponents of Mechanism certainly shared was a growing dissatisfaction, already apparent toward the end of the seventeenth century, with the explanatory model that Mechanism offered to account for the human body. The "failure" of Mechanism was that it was not able to answer all the physiological questions the body raised: in the first part of the eighteenth century, physiological research highlighted processes that could not be ascribed to a mechanical functioning of the body. For Jacques Roger, the Mechanist conception of life was destined to clash constantly with simple facts available to observation,[83] and Canguilhem is of the same opinion when he underlines the incapacity of Mechanist theories to account, for example, for gestation.[84] Mostly though, Mechanism was under attack for not being empirical enough and striving to extend principles of mechanics to the specific phenomena of the human body.[85] The critics of Mechanism, regardless of their exact indictment, were generally showing a concern with life and its unique, extraordinary functioning within the physical world—a uniqueness that Mechanism programmatically denied.

In the late 1740s and 1750s, a group of physicians from Montpellier, who had studied in Paris with Diderot, d'Holbach, and Rousseau at the Jardin Royal, started to form what is now known as the Vitalist movement.[86] Among a variety of research interests and programs they pursued in Montpellier, those who gravitated to this group shared a preoccupation with the distinctive, unique profile of features of living matter. A collaboration began between these Montpellier Vitalists and the coterie of writers involved with Diderot's *Encyclopédie*, which continued until the end of the 1760s and yielded a significant number of medical entries.[87]

In the entries drafted by the Montpellier physicians, one can trace the basic elements of Vitalism.[88] Following the project's suspension in 1759, a second generation of Montpellier doctors also collaborated on the *Ency-*

clopédie up to the point of its later publication in 1765. Jean-Jacques Rousseau played a relevant role at this juncture: the second generation of Montpellier's Vitalists, which included Henri Fouquet and Menuret de Chambaud, saw in him an ally for waging some of the principal Vitalist battles, mainly that of public hygiene, which for Rousseau played out as an indictment of city life and luxurious living.[89]

The years when the transparent-hearted human appears in the narratives I am considering are the years where these two representational systems—Mechanism and Vitalism—are vying for epistemic dominance, both striving to offer the most persuasive account of humans' true nature. It would be preposterous to impose narratives of succession and replacement on the tension and concurrence between the two systems, but Mechanism as it was practiced in the second half of the seventeenth century would eventually lose traction one hundred years later.[90]

That said, the transition from mechanistic representations of the body to vitalist models of the human was achieved, as Anne Vila has argued, between the 1740s and the 1770s, when a new image of the human being, antagonistic to the one of man-machine, emerged: that of a sensitive human, whose living functions were not reducible to those of a mechanism.[91] Vitalism shifted the representation of the body from one of a transparent mechanism to one of a sensible system, thereby making the mechanical representation of a transparent human unattainable and desirable at the same time. Far from remaining isolated within medical discourse and practice, this shift widely impacted various fields of contemporary culture. The satirical skepticism about mechanist knowledge and representations of the body epitomized by Sterne's passage therefore owed much of its edge to the growing prominence of Vitalist motives.

But what happened to the transparent body advocated by Mechanism in the age of Vitalism? As I have already mentioned, the tension between these two systems eludes the clear-cut narrative of a succession or substitution. Neither all the ideas current under Mechanism nor its larger aspirations were simply discarded. A very significant entry in the *Encyclopédie* helps us grasp the form that the tension between these two philosophies assumed. The entry "Semeiotique," devoted to semeiotics—the medical reading of signs—presents the tasks of the physician through a dialectical arrangement of mechanist and vitalist elements:

There is no part in the human body that cannot furnish some sign to the enlightened observer; all of the actions, all of the movements of this *marvelous machine* are to his eyes like so many mirrors, where the internal dispositions—be they natural or counter to nature—are reflected and depicted; he alone can direct *a penetrating gaze* into the most hidden recesses of the body, distinguish therein the state and the disorders of the various parts, recognize through external signs the illnesses that are attacking the internal organs, and determine the particular character and seat of those illnesses. It seems, judging from the easiness with which he understands what happens inside the body, that this is *a transparent machine*; but if one looks from higher above, so to say above man, the physician knowledgeable in semeiotics can look farther: the mysterious veil that hides the knowledge of the future from frail mortals tears open before him; he sees with a confident eye the different changes that must occur in health or illness; he holds the chain that connects all events, and the first links that come into his hands reveal to him the nature of those that will come after because Nature varies only in her external appearances: deep down, she is always uniform, and always follows the same course.[92]

In this passage, the complexity of which makes it necessary to quote extensively, the semiotic analyst—i.e., the skilled physician—achieves a vision of the body as transparent by unfolding the body's hidden creases, dispelling its opacity through the abstract operations that lead the physician to visualize a transparent machine. Crucially, this transparency has been achieved as a form of strenuous and skillful abstraction, since the body, of course, is not transparent at all. But the most important announcement the passage delivers is that the aspirations and capabilities of the physician far surpass those of a mechanist yearning to attain a transparent model. By undertaking a true bird's-eye view, the physician won't see the transparent body any longer, nor will he need to. His focus has changed: he sees the disease within a much broader chronotopic map that discloses chains of causation and reveals prognostic outcomes. In other words, the transparent body is neither the main nor the ultimate focus of the medical gaze and is destined to be overcome by a higher, broader, far richer vision: a vision from above, a bird's-eye view of nature as a whole landscape. Still within a highly visual paradigm of medicine (after all, the entry is devoted to the observation of signs), the image of a transparent human is now exposed as insufficient, a soothing but ultimately underwhelming aspiration.

Vitalism generated and developed further ideas that made their way to the anthropologies embraced by Sterne and Rousseau respectively. In particular, it elaborated on the notion of sensibility, which provided a significant "missing link" between the body and the mind.[93] The notion of sensibility, as I will outline it in the next section, urged a radically alternative model to that of a mechanic (and transparent) body—one aligned with opacity and one that was to become extremely productive in the fields of aesthetics and literature.

Sensibility Between Medicine and Poetics

The range of sensibility's fields of application, from medicine to aesthetics to ethics, made up a great part of the Enlightenment's intellectual output.[94] Because a discussion of this pervasive notion exceeds the scope of this research, I will not attempt to summarize it here.[95] Instead, I will follow one specific genealogy of the notion, among those extensively expounded by intellectual history, that helps best explain Jean-Jacques Rousseau's notion of sensibility and clarify the epistemic-aesthetic junction with which I am dealing. Since representations of the body and their metonymic offshoots such as the figure of the transparent heart are at the center here, it is fitting to look at physiological and medical discourse first when tracing sensibility's genealogy. As it turns out, this strain of discourse is one of the main threads (and in relation to Rousseau the most prominent) that form the composite notion of sensibility, which results from the encounter of "vitalist medicine, sensationist epistemology, moral sense theory, and aesthetics, including the novel of sensibility."[96]

In his 1752 treatise *Dissertation on the Sensible and Irritable Parts of Animals*, Swiss scientist Albrecht von Haller (1708–77), student of Dutch doctor Herman Boerhaave, laid the foundations for the subsequent dissemination of the notion of sensibility that pervaded Europe in the eighteenth century.[97] Against Boerhaave's theory of muscular contraction, which implied the passivity of the muscle moved by the nerves, Haller, through his experiments, advanced the theory of muscular fiber as the source of movement: irritability was the capacity to contract, independently from one's will. This shift carried huge implications, ultimately a remapping of the body through "a decentralized topology of sensible and irritable fibers and organs."[98] In his medical dissertation from 1752, Haller drew a crucial distinction between irritability and another property, which would soon become discursively prominent: that of sensibility, which defined the reaction of pain induced by a certain stimulation of bodily parts. Haller defined both properties: "I call the irritable part of the human body that part which becomes shorter when any foreign body touches it somewhat forcefully . . . ; I call the sensitive fiber of man that which, when touched, transmits the impression of this contact to the soul."[99] Sensibility, measured by degrees of pain, was a measure of nervous stimulation. Irritability and sensibility were therefore distinct properties, the first referring to movement alone, the second to pain; the first independent from the soul and will, the second a manifestation of the soul.[100]

Theories of sensibility within physiological discourses multiplied. Théophile de Bordeu (1722–76), perhaps the most prominent figure of the Montpellier School of medicine and a major contributor to the *Encyclopédie*, elaborated a notion of sensibility that differed substantially from Haller's.[101] For Bordeu, *sensibilité* was the ability of the body to respond to stimulations of the nerves by means of an adapted response of the implicated organs. The sum of the responses of single organs came to constitute what Bordeu and subsequent doctors designated the "animal economy," which was epitomized by Bordeu's likening of the body to a swarm of bees, which we saw when analyzing Sterne's "dioptrical bee-hive."[102]

The very same Théophile de Bordeu appears as one of the protagonists of Denis Diderot's fictional dialogue *Le Rêve de d'Alembert*, written in 1769 and published posthumously in 1830, which features Bordeu and Mademoiselle d'Espinasse conversing in the vicinity of a half-sleeping d'Alembert. The dialogic work is sustained by a playful equivalence under the sign of dream between medical theory and philosophical musings, as the exclamation of a confused Madame d'Espinasse makes explicit: "I can now proclaim to all the world that there is no difference at all between a doctor awake and a philosopher dreaming."[103]

In this work, the character of the Vitalist Bordeu exposes a theory of sensibility as the fundamental vital property; he identifies "a sensitive, living molecule" (*la molécule sensible et vivante*)[104] as the unit of life as such. In the dialogue, the physiological system outlined by Bordeu (who acts as the spokesperson of Diderot) is based on the speculative existence of a raw, primary substance of life: sensibility ("life and sensitivity . . . almost identical qualities"; "deux qualités presque identiques; la vie est de l'agrégat, la sensibilité est de l'élément").[105]

Sensibility is organized in a network, or web (*réseau*), of threads (*filsbrins*) structured in bundles (*faisceaux*) that communicate sensations from the periphery of the body to its center (i.e., the brain). The ontology that proceeds from this physiology is starkly anti-Cartesian in its repudiation of dualism: there is only one matter, endowed with sensibility, and structured with variable degrees of organization.[106] The way sensibility is described here is decidedly nonvisual, and sensation (as an effect of sensibility) is defined through nonvisual properties:

If this great diversity of tactile sensations did not exist, we should know we were feeling pleasure or pain, but should have no idea what to relate them to. We should have to rely on sight, and that would no longer be a matter of sensation, but of experience and observation.[107]

While this nonvisual characterization of sensibility doesn't imply the complete abandonment of the highly ocularcentric paradigm of the body-machine, it certainly poses that paradigm as insufficient and problematic.[108] This insufficiency is explicitly thematized by Vitalists, as in the *Encyclopédie*, in the entry "Œconomie Animale," written by another key figure of Montpellier Vitalism, Menuret de Chambaud (1739–1815), who traces how the motives of Vitalism were grafted onto Mechanism precisely through sensibility:

The human body is a machine of the kind of those called static-hydraulic, composed of solids and fluids, its first elements, common to plants and animals, being living atoms, or organic molecules: let's picture the marvelous assemblage of these molecules . . . but without *life*; this machine formed in this way doesn't differ from the living man but for movement and sentiment, the main phenomena of life probably reducible to one single primitive phenomenon; one can observe, even before life begins, or short after it has ceased, a singular property, the source of movement and sentiment attached to the *organic* nature of the principles that compose the body, or rather depending on *that* union between the molecules, which Glisson has first discovered and called *irritability*, which is actually nothing else but a mode of sensibility.[109]

The machine described in the first part of the passage—the human machine of Mechanism—is devoid of life. To speak of life, the focus has to shift from the mechanistic description of the body to the phenomena of "movement" and "sentiment," both of which depend on the underlying and more fundamental phenomenon of sensibility (conveniently indexed as a further reference at the end of the entry). The notion of sensibility emphasized here approximates that of the unique finalistic organization of the living being as opposed to the machine and its extrinsic ends.[110]

The process through which sensibility thickens and spreads as a notion across discursive fields from physiology to epistemology to ethics to aesthetics is neither unidirectional nor deterministic. To complicate the process is the semantic landscape of sensibility, a lexical cluster that includes several notions that are related, synonymous or translations, all hinging on the semantic nucleus of the Latin *sentire.* How exactly the notion migrates from medicine to other epistemic fields, and ultimately how it intersects with literature, is a question that cannot be conclusively settled. I will limit myself to show one more textual place where the discursive juncture at the crossroads of medicine, morals, and arguably aesthetics is already at play and highlights a specific representation of the body that challenges the premises of mechanistic transparency analyzed earlier.

In the *Encyclopédie*, "sensibility" appears under the rubric of medicine, as well as that of morals. The medical entry, written by the Montpellier physician Henri Fouquet (1727–1806), defines sensibility as follows:

> SENSIBILITY, SENTIMENT (Medicine). The faculty of feeling, the sensitive principle, or the sentiment of the parts, the basis and conservative agent of life, animality par excellence, the most beautiful, the most singular phenomenon of nature, etc.
>
> Sensibility is in the living body a property that some parts have to perceive the impressions left by external objects, and to consequently produce movements in proportion to the degree of intensity of this perception.[111]

The definition, which goes on at length beyond the quotation, poses sensibility as "animality as such," that is, the ultimate vital property and a true product of Vitalist thinking.

The second article on sensibility in the *Encyclopédie*, written by Louis de Jaucourt (1704–79), addresses it in moral terms:

> SENSIBILITY (Morals). Delicate and tender disposition of the soul that makes it easily moved, touched. The sensibility of the soul, as the author of *Les Moeurs* accurately puts it, imparts a kind of wisdom about propriety, and it goes farther than the penetration of the mind alone. Exuberance may prompt sensitive souls to make mistakes that men of reason would never commit; but they gain so much through the abundance of goodness they generate. Sensible souls get more out of life than others; good and bad multiply to their benefit. Reflection can make a man of honor; but sensibility makes a man virtuous. Sensibility is the mother of humanity and of generosity; it increases worth, it helps the spirit, and it incites persuasion.[112]

In its moral definition, sensibility allows the soul to be "moved" and "touched," actions that result in peculiar moral dispositions. These dispositions are described as adding a dimension of existence to those who are naturally provided with them, as the text seems to imply. Sensibility grants "more of existence than the others have"—a claim that is certainly less socially innocuous than it may seem, involving an inborn endowment with a property that grants a superior moral setup and existential capital.[113]

The strong interconnectedness of the physical and the moral notions established in the *Encyclopédie* is not just a matter of contextual polysemy; it permeates the semantics of sensibility in each of its contexts of use. Speaking of sensibility at the time of the *Encyclopédie* entries means already mapping it onto a polysemous network, from where the notion finds its way into its aesthetic declensions. Another look at the aforementioned *Le Rêve de D'Alembert* shows this further:

> But what is a sensitive being? One who is a prey to the vagaries of his diaphragm. If a touching word strikes his ear or a strange sight his eye, then at once he is thrown into an inner tumult, every thread in the bundle is stimulated, a shudder runs through him, he is overcome with horror, his tears begin to flow, he is choked with sobs, his voice fails him, and in fact the central point of the network doesn't know what is happening to it; all calm, reason, judgment, instinct, resourcefulness have fled.[114]

The character of Bordeu, here spokesperson of Diderot, believes in sensibility as the primary vitalist principle but also disparages an overblown cult of moral sensibility; in the passage, he opens up the moral definition of sensibility given in the *Encyclopédie* toward a further aesthetic deflection, here described in emphatic and highly critical terms.[115]

Outlining this specific genealogy of sensibility against other possibilities is meant to pave the way for a fuller understanding of Rousseau's use of the notion, especially in relation to the figure of the transparent heart.[116] As much as Rousseau's confessional poetics has been thoroughly discussed, and his penchant for the trope of transparency extensively expounded, notably by Jean Starobinsky's classic study *Jean-Jacques Rousseau, Transparency and Obstruction* (1957), a more definite connection of this trope with the physiological genealogy of sensibility is still lacking. In the following section, I will try to fill in this void.

Rousseau and the Poetics of the Transparent Heart

A SENSIBLE HEART

Rousseau's *Confessions* opens with a declaration of exceptionality: "I am resolved on an undertaking that has no model and will have no imitator. I want to show my fellow-men a man in all the truth of nature; and this man is to be myself."[117] The uniqueness of the enterprise rests on the uniqueness of the subject—that is, the uniqueness of Jean-Jacques Rousseau's narrator—as is concisely expounded in the second paragraph of the book: "I feel my heart and I know men. I am not made like any that I have seen; I venture to believe that I was not made like any that exist. If I am not more deserving, at least I am different. As to whether nature did well or ill to break the mould in which I was cast, that is something no one can judge until after they have read me."[118] The narrator bases the claim of his unrivaled singularity on two distinct and interrelated properties: the feeling for his own heart (*je sens mon cœur*), and his knowledge of human beings (*je connais les hommes*). As far as the latter anthropological claim goes, Rousseau put as an opening to his work a *captatio benevolentiae* where he defined his work as unique and useful, which could serve as the cornerstone for an as-yet-unwritten study of "man."[119] The opening claim of the first book—the full disclosure of the narrator's nature—is established on the basis of this double foundation: sensibility and anthropological knowledge.

In the *Confessions*, soon after the early account of his birth, Rousseau defines his *cœur sensible* as an inheritance from his parents: "Of all the gifts bestowed on them by heaven, the only one they bequeathed to me was a tender heart; but to this they owed their happiness, just as I owe it all my misfortune."[120] The legacy of a sensible heart is here blamed as the main source of unhappiness in Rousseau's life, therefore at the same time the source of his unique ability for self-revelation and of his unique suffering. Rousseau clearly embraced the Encyclopedists' account of a double sensibility: physical and moral, both of which he believed he possessed in great amounts.

For Vila, it is on the scientific ground we have reviewed in the section above that Rousseau could implement his *morale sensitive*—his sensibility-based ethical code.[121] What is certainly emerging from the passages of the *Confessions* just quoted is that Rousseau invoked the physiological capacity to feel, bestowed by inheritance, to establish a morals of sensibility that was intended as a marker of uniqueness. Adopting the general theory of sensibility meant, for him, bending it toward an idiosyncratic, personalized version that, in the context of his confessional writing, also served as a self-apology in view of a confessional pact with the public.

The early characterization of the narrator's sensibility in the *Confessions* is carried on with reiterative insistence: he declares his sensibility to be prior to his thought, "I had feelings before I had thoughts" (*Je sentis avant de penser*), and along the same line of argument, he confesses that feelings not only preceded but replaced cognitive appraisal: "I had as yet no ideas about things, but already I knew every feeling" (*Je n'avais rien conçu, j'avais tout senti*).[122] Rousseau's invocation of sensibility as prior to other forms of cognition is not limited to the account of his own. In his pedagogical work *Émile* (1762), he declares: "To exist is to feel; our feeling [*sensibilité*] undoubtedly precedes our intelligence."[123]

In the third book of the *Confessions*, Rousseau describes the discrepancy generated by his excessive endowment with sensibility compared to his cognitive capabilities in even starker terms as a conflict between the fervor of his sentiment (*cœur*) and the slowness of his thought (*esprit*):

> Two almost irreconcilable opposites are united within me, without my understanding how this can be so: on the one hand, an ardent temperament, keen and impetuous passions, and on the other, ideas that are confused, slow to take shape, and only ever occur to me afterwards. It is as though my heart and my mind belonged to different people. Feelings burst upon me like lightning and fill my soul; but instead of illuminating, they burn and dazzle me. *I feel everything and I see nothing.* I am transported yet stupid; to be able to think, I need to be composed. The surprising thing is that I am capable, nonetheless, of sureness of touch, penetration, subtlety even, so long as I am not hurried; my impromptus, given time, are excellent; but I have never, on the spur of the moment, said or done anything worthwhile.[124]

In the passage, the effect of "sentiment" is described as a blinding force, and an antithesis between feeling and sight is established: "I feel all, but see nothing" (*Je sens tout et je ne vois rien*). Given the mutually exclusive forces portrayed here (feeling as opposed to seeing, sensations as opposed to cognitions), it is even more striking that Rousseau, in his quest for self-representation, resorted to a very different image than that of a sensible heart: the image of his heart/soul as transparent, perfectly visible, unashamedly exposed to the view of the reader.

As we will see in the paragraphs to follow, sensibility remained the main foundation of Rousseau's morals and anthropology, with the paradoxical consequence that this Vitalist notion was fundamentally at odds with the epistemic aspirations of confessional disclosure advanced in the *Confessions*. The image of Jean-Jacques's "transparent heart" lets these tensions converge, ultimately functioning as a critique of the epistemic tenets of Enlightenment and, therefore, of eighteenth-century humanist discourse.

A TRANSPARENT HEART

The unique and exceptional qualities of Jean-Jacques are allocated to his heart: "a heart too affectionate, too loving, too tender, which for want of real people of its own kind is forced to feed on fictions."[125] In Rousseau's work, the heart is a pervasive organ, symbolically carrying out a number of functions that relate to sensibility. The heart can be respectively the site of the soul, of emotions, of morality—a truly powerful body part. While cardiac physiology of his time had certainly dissociated cognitive functions from the work of the heart (which had been a tenet of ancient physiology),[126] most of Rousseau's references to the heart can be read in terms of Ernst Robert Curtius's "corporal metaphors," a wide family of *translata*, which, from at least the Bible onwards, confer spiritual properties to certain parts of the body.[127]

Eighteenth-century culture attests to an extensive recourse to the language of the heart or *eloquentia cordis*, which was revamped in Pietism and had its roots in biblical topoi.[128] Still symbolically rooted in a proper physiology of the heart (Hippocratic but also Aristotelian) prolonged in topical Gospel passages,[129] the language of the heart accessible to Rousseau was a rhetorical precipitate of discursive layers that retained symbolic force more than explicative validity. Rousseau's "heart" is a corporal metaphor or, more precisely, a metonymy or a synecdoche, whether it is evoked to express properties allegedly deriving from it or to symbolize the human being as such. It is in the first way, for example, that in *Émile* the heart is postulated as the site of the moral life of the human: "Let us lay it down as an incontrovertible rule that the first impulses of nature are always right; there is no original sin in the human heart, there is not a single vice whose infiltration cannot be explained and traced."[130]

In a consistent number of cases, the heart evoked by Rousseau as his own is made of glass, or crystal, or is simply described as transparent, as in a further confessional appendix to his *Confessions, Rousseau juge de Jean-Jacques*: "His heart, transparent as crystal, can hide nothing of what goes on within it; every emotion he feels is reflected in his eyes and on his face."[131]

If the image of the transparent heart has crystallized as a recognizable feature of Rousseau's poetics, it is in large part owing to Jean Starobinski's comprehensive hermeneutic reading of Rousseau's work. Starobinski's critical merit, in his seminal book dedicated to Rousseau, *Jean-Jacques Rousseau, Transparency and Obstruction*, has been to gather a variety of semantic fields perused by Rousseau under the capacious, visually captivating, and polysemous image of transparency. Starobinski grasped the efficacy of a visual expression, such as that of transparency, over a more abstract notion like immediacy, which on closer inspection seems to be the actual matrix of all the semantic fields gathered by the critic under the label of transparency. Whereas the notion of transparency, as such, is not ubiquitous in Rousseau's work, his writings are filled with images of visibility and concealment—for example, when he opposes the dynamic of concealment that marks civilization to the state of unashamed exposure characteristic of what he imagines as primitive humanity. Already in the *Discourse on the Arts and Sciences* (1750), Rousseau had employed a scopic metaphor to describe social relations: "At base, human nature was no better, but men found their safety in the ease with which they saw through each other."[132] For Rousseau, visibility as exposure is the natural state of existence, and it defines a condition of transparency, accordingly called by Starobinski "primal transparency" or "primordial transparency."[133] The image of a transparent heart is the image that Starobinski places at the center of Rousseauian poetics, promoting transparency to master-trope within a system of variations. A transparent heart recurs consistently throughout Rousseau's work, up to his late correspondence. As Starobinski reminds us, Rousseau used to open his letters with the lines "What poor, blind creatures we are! / Heaven, unmask the impostors / And force their barbarous hearts / To open to the gaze of man."[134]

In the remarkable hermeneutical thrust to bring together every textual element from Rousseau's life and work, Starobinski finds in the *Institutions chimiques*, particularly those passages devoted to the chemical process of vitrification, the material source of meaning for Rousseau's notion of transparency and, at the same time, the material coincidence of transparency and immediacy. The mention, in Rousseau's writings on chemistry, of "water and liquids among whose parts their transparency shows an immediate union," leads Starobinski to conclude that "in the physical world immediacy and transparency are correlative notions. Light can pass through certain substances because those substances possess the perfection of immediacy. The hypothesis may be 'chemical,' but it expresses a psychological necessity."[135] For Starobinski, fluidity is therefore identified by Rousseau himself as the physical condition of transparency, and on this basis the critic, with a sweeping retrospective gesture illuminating his whole study, concludes that "if fluidity is the cause of transparency, the relation between the metaphors 'crystal' and 'clear water' is closer than we might have thought. In both cases, it is inner unity that permits light to pass."[136]

While Starobinski's hermeneutic analysis remains a milestone of rhetorical analysis and literary criticism, my interest in the image of the transparent heart departs from this analysis in pursuing a narrower goal. As a result, where Starobinski finds homogeneity in the choice of images and distills their common denominator in the all-encompassing property of transparency, my aim is to show a dissonance in the imagery employed by Rousseau. Where Starobinski is putting the pieces together and finding their common denominator, I am searching for the missing link—not in order to insert it back but to emphasize the disruption in Rousseau's system of thought.

The main point of friction that I want to show here pertains to the language of sensibility and that of transparency. As we have seen in the previous paragraphs, sensibility sprang from medical experiments focused on the nervous system and was defined either by the reaction of pain in the body (Haller) or by the coordinated system of perceptions between parts of the body (Bordeu and Fouquet). The *Encyclopédie* defined sensibility in its medical and moral meaning as the property of existing at a different depth, overcoming the visual boundaries of understanding, something that both a good doctor (in the entry "Semeiotique") and a virtuous man (as stated in the moral entry "Sensibility") shared. In other words, sensibility was a nonvisual property, defined variously by tactile and nervous functions of the body or by emotional disposition. The Swiss physician Albrecht von Haller had conducted gruesome vivisections, inflicting pain reactions on animals, to arrive at his notion of sensibility.[137] Outside the torture room of the anatomic theater, however, sensibility was not measurable. Claiming one's sensibility was not easily proven, and this might explain why Rousseau was so insistent in his autobiography about his owning one in excessive amounts. For somebody obsessed with the public reception of his persona, however, insisting on his sensibility was not necessarily a successful strategy, as there were no concrete measurable ways to demonstrate his claims. The public would ultimately want to witness Rousseau's innocence and virtue incontestably, and with their own eyes: Rousseau needed to show something, to make himself visible. His constant concern with disclosing himself to his readership was linked to a pervasive fear of accusation:

> But the undertaking I have embarked on, to reveal myself to it [the public] in my entirety, requires that nothing about me should remain hidden or obscure; I must be continually present to its gaze; it must follow me into all the aberrations of my heart, into every recess of my life; it must not lose sight of me for a moment, for fear that, finding in my story the least lacuna, the least void, and wondering to itself what I did during that time, it should accuse me of not wanting to reveal every-

thing. I am giving enough quarter to the malice of men by telling my story without giving even more by remaining silent.[138]

From the perspective of this concern with publicity, it is easy to see the trope of transparency being deployed as a rhetorical device intended to guard the confessant from the confessor's mistrust and disbelief about deceptive, disingenuous exposure. Rousseau needed to show his uniqueness—that is, to show his unique sensibility, to make apparent and visible something that was intrinsically invisible.[139]

In the passage from the *Confessions* that I have analyzed at the end of the previous paragraph, sensibility was described as a groping dazzle and inescapable blindness: "Feelings burst upon me like lightning and fill my soul; but instead of illuminating, they burn and dazzle me. *I feel everything and I see nothing.*"[140] The function of the image of the transparent heart was supposed to locate in the heart the property of sensibility in order to make it visible with the help of the well-known device of a glass of Momus, to which Mechanism had granted epistemic and technical plausibility. The struggle for Rousseau, however, was real, and a metaphor alone couldn't protect him from his fears. This is why, in the passage quoted above, he stretched the metaphor of the transparent heart toward the field of physiognomics: "His heart, transparent as crystal, cannot hide anything that happened; every movement he feels is transmitted to his eyes and face."[141] It is precisely the epistemic ambition declared in this passage that may have sifted layers of signification to then settle in the heart of Jakob W***r, whose tale was told at the beginning of this chapter. If the *Encyclopédie* defined sensibility as what makes the soul be touched and moved, physiognomy mapped the "movements" of sensibility onto facial features, making sensibility visible. Physiognomy established the most superficial (in the literal sense of surface) link possible between the body and the soul. The personal doctor of Louis XIV and fervid sustainer of physiognomy before Lavater,[142] Marin Cureau de La Chambre (1594–1669), in his treatise *L'art de connoistre les hommes* (1660) had written:

He was wrong, that man who complained once that Nature hadn't placed *a window on the heart* so that one could see thoughts and plans of humans. Not only these are things that the senses cannot perceive, and even assuming that the eyes could actually see the bottom and the inner folds of the heart, they would not draw any knowledge from that. But even more than that, Nature actually provided the means to discover all this, and means that are much more reliable than this strange opening that Momus had imagined.[143]

For La Chambre, the human face revealed the thoughts and intentions of its owner, and it did so mechanically: every muscle moved in a certain way when reacting to certain drives. While evoking and dismissing as "strange" the idea of a window on the heart, La Chambre conceived of physiognomy as a plastic, mechanical representation of the soul writing itself on the human face.[144]

The epistemological concerns of physiognomy, in the first place its preoccupation with unmasking deceit, were deeply grounded in the cultural landscape of the late seventeenth and eighteenth centuries.[145] Rousseau drew from physiognomy's traditional toolbox to uncover mystification and deceit by proposing, however, something entirely different from facial interpretation. For him, the face was illegible, since it had already been disfigured by social conventions. To find the supposed truth of human nature, it was necessary to cut deeper (*intus et in cute*, as the epigraph to the *Confessions*, a verse from a satire by Roman poet Perseus, pleads)—indeed, to reinstall a window on the heart. Through this claim, Rousseau was advocating a mechanic representation to accomplish a task that had largely exceeded Mechanism's field of interests, namely the representation of the soul. In so doing, Rousseau was juggling conflicting systems and incompatible imageries in the attempt to build his own idiosyncratic poetics and morals—something he, indeed, managed to do, albeit at the cost of exclusion and alienation from more than one intellectual circle (be it the French *philosophes*, the *Encyclopédistes*, the English entourage surrounding Hume, and so on).[146]

The transparent body had now been in vogue as a mechanistic model of the human body for at least a full century, and only recently had this picture started to be contested by the advocates of Vitalism. Rousseau had intercepted the new wave and played a major part in the dissemination and reproduction of one of its mainstays—the notion of sensibility—but he was also bound to the visual paradigm championed by Mechanism for reasons that included fashion, intellectual credibility, and paranoia. Unlike what Starobinski strived to prove, there was no overarching consistency or homogeneity in Rousseau's appeal to the transparency of the heart but, rather, a strenuous endeavor to legitimize his autobiography and publicize his persona while holding on to a picture of himself that had nothing visual about it: that of the unrepresentable, sensible soul. The appeal to the trusted image of Momus's glass could shelter his self-representation from anticipated accusations and resonate with the public as something more than the mere repetition of an ancient myth: it offered the possibility to reintegrate mechanistic fantasies and vitalist aspirations.

Rousseau's image of the transparent heart is therefore the dissonant convergence of medical debates of the eighteenth century, of his own entanglements with the construction of his literary persona, and of the rhetorical sediments of a long-lasting "language of the heart." Through the image of the transparent heart, Rousseau was revamping the trope of Momus's glass by reinforcing it with the epistemic ambitions of Mechanism and injecting it with the vocabulary of sensibility. While establishing what would retrospectively become a paradigm for autobiographical, confessional exposure, he was delivering through a transparent heart a critical exposure of the shortcomings of Enlightenment episteme.

A Posthuman Device

Looking for the cultural historical point of convergence where the image of the transparent heart in the contemporaneous narratives of Sterne, Rousseau, and Spieß intersect, this chapter has traced it back, through its entanglements with traditional and newly reinscribed tropes, to the medical episteme of the eighteenth century. The middle years of the eighteenth century had seen the consolidation of Vitalist medical research against the medical discourse that had been prevalent in the seventeenth century with a long coda into the eighteenth: Mechanism. Mechanist medicine had relied on the assumption that the body was perfectly decipherable and understandable if considered as a machine and had provided accounts and technical models of the physiology of the body based on the postulate of its transparency. In the anatomical-physiological model of the German physician Salomon Reisel (1680), I have identified a visual epitome of the Mechanist representations of the body. The *Statua humana circulatoria* was a machine working through siphons, reproducing the main physiological processes of human life in a simplified way—its heart and its brain designed to be of glass (the brain a glass sphere with the pineal gland as a crystal hanging within).

When Sterne, in *Tristram Shandy*, criticized the flaws and pitfalls of Mechanism, he used the image of a transparent heart to ultimately dismiss the pretense of a mechanistic knowledge of the human being, advocating for its baffling, elusory nature, which could be visualized, if at all, only in terms of organic impenetrability. On the contrary, when Rousseau invoked at several points of his work the transparent quality of his heart, he was championing the possibility of knowing the human being as the myth of Momus imagined it: seeing through the soul clearly, without deception. But

Rousseau was at the same time promoting an anthropology and morals of sensibility as a means to establish his personal and authorial uniqueness. Sensibility, a scientific offshoot of Vitalist medical research, was at odds with the picture of a transparent human. Based on the research on irritability and pain, sensibility conceived the human body as an invisible network of sensations and feelings. Rousseau's insistence on the perfect, mechanistic visibility of his own sensibility was therefore a highly oxymoronic gesture, signaling a tension both within his own aesthetic system and in the broader culture of the Enlightenment.

The figure of the transparent human delivered by the narratives of Sterne and Rousseau, in different ways, is one that approximates its obsolescence. Proclaiming its aspiration to grasp, capture, and draw a model of the human being, this figure cannot withstand the affront of sensibility and of the Vitalist, nonvisual representation of the human. The mobilization of the image of the transparent heart in two of the most significant literary autobiographies of the eighteenth century signals a fundamental negotiation at play between concurrent medical and broader epistemic representations of the human being, with their repercussions on the literary imagination. Both autobiographical texts expose the visual paradigm of the scientific rhetoric of the Enlightenment together with its shortcomings: the transparency of the heart would fail (either as impossible or as insufficient) where the knowledge of the human was the knowledge of a particular individual. The autobiographical task of rendering the life of the single individual posed an insurmountable challenge to the anthropological project of the Enlightenment, one that Rousseau thought could succeed also thanks to his confessional endeavor.[147]

But both these narratives, through the topical recourse to the figure of the transparent heart, make tangible how the transparent human had become a chimera, and the only available knowledge now hinged on invisible feelings and blinding emotions, on the one hand, and the burdensome flesh of the body, on the other. It is then all the more significant that in the late Enlightenment case-study of Jakob W***r, written by Hans Christian Spieß some years after Sterne's and Rousseau's narratives, the transparent heart returns as a figure of pathological delusion, reevoking the melancholic complex of early modern culture. The transparent heart as an imagined prosthetic device allowing complete visibility and knowledge of the inner life of a human being, while endorsing a hyperhumanist embodiment was also coming short of its own humanist premises. In Jakob W***r, the epistemic ambitions of the humanist project of the Enlightenment, which physiognomy had seized and conveniently reduced to a readily acquirable practice, come to an end as they become a threat to the protection of one's own treasured mystery and soothing opacity. Yet Jakob W***r in the short story held on to his delusion the same way that glass melancholics of early modernity had done, because it was in this pathological configuration that the source of their exceptionality could be found. In other words, the figure of a transparent heart kept operating as a promise that, although impossible to keep, could at least secure a critical function within the episteme that had produced it. A transparent heart may be the obsolete delusion of Enlightenment, but it also holds on as the metonymical stage of a posthuman dream within the age of humanism.

THREE

A Human in Glass

GOETHE'S HOMUNCULUS AND ROMANTIC EMBODIMENT

The metonymical glass human of the Enlightenment (a human with a transparent, crystal heart) gives way, in the early decades of the European nineteenth century, to a further metonymical fantasy with similar epistemic force yet directed toward very different epistemic ends. The tale of Homunculus told in Goethe's *Faust II* (1832) showcases a glass human very different from the ones we have seen so far, while nonetheless sharing with its antecedents a fundamental epistemic interrogation of the conditions of human embodiment. As a humanoid creature born into glass, and provided by Goethe with the firm intention of shattering that glass and becoming fully human, Homunculus displays, in this master narrative of Romantic episteme, a reverse transition from vitreous to fleshly embodiment, spelling backwards, as it were, the swapping of flesh with glass, thereby challenging some of the mainstays of humanism from a different angle: showing alternative paths to becoming human, as well as their tremendous costs.

In this chapter, I will examine why this miniature character is a momentous figure in the cultural history of glass bodies by discussing the cultural negotiations entertained by this historical hybrid—part history of alchemy, part folklore—alongside the cultural references mobilized by Goethe in *Faust II*. Although Goethe's contribution to Romantic poetics is as unquestionable as it is contested, through *Faust II*, and in particular through the tale of Homunculus, Goethe delivers a thick narrative of Romanticism's epistemic aspirations and their failings. Within this framework, I read Homunculus as a *sui generis* glass human—a being whose material and bodily shifts are to be interpreted within the cultural historical complex of what I will call Romantic embodiment, with all its intrinsic tensions.

Homunculus appears in the second act of the tragedy's second part, and then accompanies Faust on his immersive travels in time and space to classical Greece. From the laboratory where he first appears to the Aegean Sea where he ultimately disappears, Homunculus lives a brief life on the stage, but his tale constitutes a hallmark of the play, leading Faust to the fulfillment of his own deepest wishes and containing in his own subplot some of the essential threads of the whole tragedy. Homunculus's existence is bound throughout to the glass vessel that constitutes the artificial womb of its alchemic creation, which allows this miniature being a unique mode of existence: ingenious, shrewd, clairvoyant. But the price this diminutive character has to pay for his powerful gift is his confinement to the glass of the vial and, even more importantly, his lack of a body thoroughly human, for which he is ready to sacrifice his extraordinary existence in the bottle.

Homunculus therefore represents a peculiar incarnation of a transparent human—a fantasy of the human charged by multiple, simultaneous experiments that entail his vitreous existence as much as his peculiar generation. As a humanoid creature, his humanity is conceived of as incomplete (which is reflected in the designation he receives of "Männlein" [little man] and in his self-perception). Furthermore, his transparency showcases a medial status that differs from the one shared by the other transparent humans we have encountered so far. The transparency of Homunculus—whose body, although endowed with a certain transparency, is not made of glass—is a function of the bottle (a chemical vial) that constitutes his medium to the world; a condition of his own embodiment, the glass bottle is the prerequisite of Homunculus's physical existence, constituting, so to say, his body. While his actual body is half-transparent (delicately translucent and shining), it is his peculiar embodiment in glass that bestows on Homunculus a transparent vision that, in fact, associates him productively with the other transparent humans I have discussed.[1]

A key figure in the complex plot of the second part of the tragedy, Homunculus can only be understood within the larger framework of the dramatic and aesthetic dimensions of Goethe's oeuvre, which is why this chapter will also dwell on the text and on the discussions around it. As with any figure in Goethe's work, Homunculus has inspired a vast repertoire of interpretive attempts to define its dramatic function within the general design of the tragedy, in particular vis-à-vis the main characters of Faust and Helena.[2] Beyond mapping the constellation of characters, cultural historical readings of Homunculus have assigned it to its proper context of origin, that of alchemy, with which Goethe was deeply familiar and where Homunculus, as such, was first conceived.

As a laboratory figure rooted in the alchemic tradition, Homunculus doubles its experimental function in Goethe's play by performing a master experiment of embodiment: assuming an actual body of flesh. Although Goethe elaborates a subtle and complex transition from artificial life to natural life, he leaves the positive description of Homunculus's new, fleshly life untold; Homunculus will disappear from the stage a moment before having broken the glass vial against the waves of the Aegean Sea, his transformation being eventually announced, in the future tense, as forthcoming. What matters most, then, in Homunculus's tale, is not so much what the completed process of embodiment in flesh signifies but rather the meaning of the transition he yearned for throughout his brief life onstage.

In proposing a specific reading of this figure within the arc of the story told in this book, I will select the clues that render this reading meaningful, sifting through a thick history of interpretation. Linked to the alchemical-historical interpretation is, for example, a strand of interpretations that descends from Carl Gustav Jung's reading of *Faust* that read Homunculus's tale in psychological and symbolic ways.[3] But the highly suggestive figure of Homunculus has further inspired readings of it from the standpoint of the theory of evolution,[4] of human cloning,[5] and, even more ambitiously, in light of models of Artificial Intelligence.[6]

In the course of this chapter, I will take into account a number of indications coming from such a diverse spectrum of critical interests to eventually pinpoint the nature of Homunculus as a transparent human fulfilling that essential function of the fantasies of the human that I have defined so far: the critical rethinking of the conditions of being human through the fictional experiment of replacing a human body of flesh with a transparent, glass body.

A Human Being's Being Made: *Homunculus Enters the Stage*

Wagner (*at the hearth*).

The terrifying bell reverberates
and sends a tremor through these soot-black walls.
The end has come to the uncertainties

attendant on my solemn hopes.
The shadows have begun to be less dark,
and in the inmost vial
something is glowing like a living ember and,
like a glorious carbuncle,
irradiates the darkness with red lightning flashes.
A clear, white light can now be seen!
If only, this time, I don't lose it![7]

The appearance of Homunculus in Goethe's *Faust II* is described as an event of light occurring within the space of the glass vial, jointly held by the alchemist and Faust's former servant, Wagner, at work in his laboratory. Act 2 of *Faust II* opens within this space: a medieval-style laboratory with bulky and extravagant tools (the stage directions specify a "medieval alchemist's chamber filled with cumbersome apparatus designed for various fantastic purposes").[8] Here, in front of the alchemist's furnace, Wagner presents to the public the contents of the vial, in the verses just quoted. The descriptive clues—"the shadows have begun to fade," "something is glowing like a living ember," "like a glorious carbuncle," "irradiates the darkness with red lightning flashes," "a clear, white light can now be seen"—all contribute to the widest possible dissemination of elements from the semantic field of light; as numerous as insistent, the resultant *chiaroscuro* effects already mark the powerful visual semiotics generated by the experiment in the vial.

After pondering the multiple phenomena of light produced by his experiment, in the scene, Wagner welcomes Mephistopheles into the laboratory and reveals to him the nature of the event transpiring in the glass:

[Wagner]
Something tremendous is just about completed.
. . . .
A human being's being made.[9]

Wagner announces that he is trying to create human life artificially (*Es wird ein Mensch gemacht*), which he proceeds to explain to Mephistopheles as the act of providing humanity with a nobler origin (*höhern Ursprung*) than the one assigned by the biological mainstays of human reproduction: "God forbid! Old-fashioned procreation is something we reject as folly."[10] Once again, the birth of Homunculus is announced and prepared by luminous phenomena taking place in the dimness of the laboratory: "It brightens! See!"; "The mass is clearer!"; "It rises: flashes, there's expansion."[11]

The fabrication of Homunculus is staged to convey the dramatic action of the laboratory experiment, a setting that openly and intentionally recalls that of alchemic work: the sparks and flames of the chemical combustion performed by the alchemist, glowing through his glass equipment—a *topos* of alchemic iconography.[12] Alchemy is also conveyed through the very contrast between light and darkness, the *chiaroscuro* mentioned above, which is traditionally a central tenet of the symbology expounded by historical alchemic texts.[13] The stages of the process that lead to obtaining the philosophers' stone correspond to precise passages through an ample codified chromatic spectrum that goes from black (*nigredo*), to rainbow (the so-called peacock's tail, or *cauda pavonis*), to white (*albedo*), and finally red (*rubedo*).[14]

At the same time, the dialectical semantics of the contrast between darkness and light that dominates the entire second act, while serving the purpose of the *mise-en-scène* of alchemic experimentation, extends further to the chromatic scheme of the whole work, where Mephistopheles's presence is often signaled and accompanied by dimness and obscurity. For example, the conversation between Homunculus and Mephistopheles about whether to carry a still-sleeping Faust to the classical Greek Night of Walpurgis provides the occasion to display the dialectic of obscurity and luminosity enacted by the two figures: "you only feel at home where gloom prevails," Homunculus rebukes Mephistopheles;[15] afterward, Homunculus appoints himself to lead the way by lending his own light: "I'll go ahead and light the way."[16]

The following scene, the third of act 2 (namely, the "Klassische Walpurgis Nacht"), is punctuated by reminders of Homunculus's luminous nature, as, for example, when Mephistopheles suggests that they should meet again, after first each going his own way, at the blinking signal of Homunculus: "our little friend must let his lamp shine bright and resonate."[17] Homunculus, at Mephistopheles's proposal, responds by reminding him that he can also signal through the sound of the glass vial: "This is the way it will flash and ring."[18] Accordingly, the stage directions specify: "*The vial resounds and emits a strong light.*"[19] The visual effects of light are here complemented by the auditory reminders of Homunculus's encasement in the glass vessel. With a further contribution to building the field of light for Homunculus, Mephistopheles exclaims, once they meet again in the course of the classical Walpurgis Nacht: "an unpretentious light is moving. / How nicely things work out."[20] The physical limitations of Homunculus, and his desire to break free, are sustained by these important synesthetic clues, which serve to conjure up the motive of the impatient pursuit of his quest for *Entstehung* (generation), which I will return to later.

Homunculus's radiant nature has not only to do with setting (the alchemist's laboratory) or with the dramatic contrasts of the play (his opposition to the tenebrity of Mephistopheles). Rather, its light appertains to his very nature and is synonymous with his talent, what Mephistopheles calls his "Gabe."[21] This talent, which Mephistopheles propels him to demonstrate while they are still in the laboratory, consists in the ability to see through what is usually hidden—in this specific case, to read Faust's mind while he dreams.

In a passage from Eckermann's *Conversations* with Goethe from 1829, Eckermann defines Homunculus as a being that can see through things: "As a being to whom the present is perfectly clear and transparent, the Homunculus sees into the soul of the sleeping Faust."[22] While hovering over Faust and casting light on his sleeping body, as the stage directions indicate, Homunculus is able to reveal the contents of Faust's dream: a mythological scene depicting Leda bathing right before her sexual encounter with the king of the swans, Zeus—the scene, in other words, of Helena's conception. The scene itself is pervaded by Homunculus's light: Leda dips her foot "into the bright transparence" (*durchsichtige Helle*); her body is metaphorically designated as a "living flame" (*Lebensflamme*), which she cools off in "the water's pliant crystal" (*Kristall der Welle*)—a scene that echoes through its images the process of alchemical creation or at least the firing operations of a laboratory.[23]

It is as if Homunculus's vision were an extension of his own physical properties, as if his being made of light cast light onto the very images of his vision, and the transparency of his sight through the vial projected a transparent sheen onto what he saw. Homunculus's clairvoyance is a complex medial process, in which the transparency of the subject (Homunculus) transfers onto the object of his vision.[24] This peculiar mediality is not new in this book. Early modern glass humans (chapter 1) were intellectually endowed with the same crystalline sharpness of the glass of their bodies, and the case of Jakob, a man with a heart of glass at the end of the eighteenth century (chapter 2), unfolded with a similarly apparent inversion: his own fear of transparency allowed him to see through natural and human impenetrability.

At stake in the various appearances of a glass body is in large part, then, the possibility of establishing the epistemic conditions for a transparent vision of the world. In the case of Homunculus, this peculiar mediality manifests through poetic language as a permeable epistemic continuity to the objects of his vision. Mephistopheles calls Homunculus "Phantast," a fantasist, a notion that alludes not just to Homunculus's inclination to fantasies and dreams but to the transformative impact of such clairvoyant activity: Homunculus makes things transparent merely by looking at them.

Once Homunculus and Mephistopheles, who are carrying along a sleeping Faust, have departed for Greece to attend the night of Walpurgis, Mephistopheles, according to the stage directions, turns to the audience with a revelation:

> The fact is, we remain dependent on
> the creatures we ourselves have made.[25]

Mephistopheles is revealing here that his role in the fabrication of Homunculus was more than accidental. Homunculus's dependency on Mephistopheles's intervention is confirmed in the same conversation between Eckermann and Goethe quoted above, where Goethe reportedly explains to Eckermann the common nature of Homunculus and Mephistopheles as regards spiritual clarity but also states a "disadvantage" of Mephistopheles in terms of recognizing physical and moral beauty.[26] This similarity is further elaborated by Goethe as a shared demonic nature unimpeded by the obfuscations and limitations of human nature: "for such spiritual beings as this Homunculus, not yet obfuscated [*verdüstert*] and limited [*beschränkt*] by a thorough assumption of humanity, were classed with the demons, and thus there is a sort of relationship between the two."[27] Goethe here allegedly provided a number of relevant considerations for understanding Homunculus's nature: a spiritual being (*geistiges Wesen*) belonging to the ranks of demons yet whose main characteristic is partial humanity, which is neither obfuscated (*verdüstert*) nor restricted (*beschränkt*) by humanity's limitations.

In Eckermann's version of Goethe's assessment, Homunculus's transparent vision is linked to his incomplete humanity and his rather nonhuman (instead, spiritual) nature. His unimpeded vision is defined as lack of limitation, even though the play constantly reminds Homunculus of its confinement in the glass cage. That is, while a demonic explanation of Homunculus's nature emphasizes his lack of limitation, the play accentuates the limitedness of Homunculus's state, complicating the neat distinction of human and nonhuman and activating a tension that the drama purports to resolve.

In the same conversation, Eckermann asks Goethe about the apparently subordinate role of Mephistopheles in the scene, which would make it difficult to accept Mephistopheles as the creator of Homunculus: "yet I cannot help thinking that he has had a secret influence on the making of Homunculus."[28] To this objection, Goethe allegedly replies that he had considered making Mephistopheles's participation in the production of Homunculus more unequivocal by changing some wording.[29]

Homunculus retains the ambiguity of being both a Mephistophelian creation and a dialectical, dramatic counterpart to Mephistopheles (the contrast between light and darkness being one of its most prominent symptoms). Mephistopheles's contribution to Homunculus's generation also holds a further ambiguity in the dramatic action, as it is mysteriously and miraculously achieved at a distance, whereas alchemical generation in its historical characterization (as Wagner has supposedly implemented it) should be a technical process based on presumed laws of nature. Contrary to Mephistopheles's intervention, Wagner has been at work on the process for some time with determination and conviction ("A human being's being made") and plausibly considers the result his personal achievement. Homunculus would seem the product of wondrous magical intervention combined with alchemic technique.

The creation of artificial life in the alchemist's laboratory was not just theorized, however; it was also illustrated by means of concrete recipes in early modern alchemical texts. These texts were known to Goethe, who had familiarity with the corpus of early modern alchemy and its symbology, and for whom alchemy played an important role in the construction of both his scientific and his aesthetic framework. The fact that, in 1828, the chemist Friedrich Böhler had synthetized the organic compound *urea* starting from inorganic substances and that Goethe certainly had knowledge of this event through his chemist acquaintances may even indicate a starting point for Homunculus's plot.[30]

Goethe and Alchemy

In his now classic study *Goethe the Alchemist* (1952), Ronald Gray documents Goethe's early interest and involvement with alchemical texts and laboratory experiments.[31] Gray's work, even if burdened by the long-lasting historiographical fallacy that has often conflated the practice of early modern alchemy with its later occultist reinterpretations and reappropriations, effectively demonstrates alchemy's persistent role in Goethe's system, alongside several other scientific interests pursued by Goethe over his lifetime.[32]

In Goethe's time, alchemy was undergoing its first modern revival following a relative decline during the opening decades of the Enlightenment: in Germany, the 1770s and 1780s saw a peak in the publication of alchemical texts.[33] This happened in large part in conjunction with the rise of secret societies, which appropriated alchemy as a secret discourse and thereby contributed to associating alchemy to an esoteric science steeped in ancient and arcane knowledge.[34]

Goethe's alchemical interests correspondingly spread on an alchemical-chemical historical continuum that included quite a wide corpus that encompasses key figures of early modern alchemy such as Paracelsus, van Helmont, and Starkey and experiments with chemical processes. In Goethe's own account, provided in his autobiography *Dichtung und Wahrheit*, he had spent the winter of 1768 (when he was just nineteen years old) reading alchemical texts together with a Pietist friend of his mother, Fräulein von Klettenberg—an experience that sparked, in Goethe's own words, a "secret love" for alchemy that lasted in this intensity until at least 1770 but that definitely left long-lasting marks on his work.[35] Goethe recounts in his autobiography how he and von Klettenberg read Paracelsus, Basilius Valentinus, Helmont, and Starkey, and he mentions the great influence exerted on him by the hermetical text *Aurea Catena Homeri* (by Anton Joseph Kirchweger, 1723), which portrayed Nature as thoroughly interwoven: "in which nature, even though perhaps in a phantastical manner, is represented in a beautiful interlacing."[36] Together with Fräulein von Klettenberg, Goethe had even started setting up his own alchemic (that is, chemical) experiments (which entailed proper equipment such as that pictured in fig. 10) and engaged especially in the production of an alkali solution, "Kieselsaft" (*liquor silicium*), also known as water glass.[37]

Goethe's subsequent interests in the natural world veered in the years around 1775 toward the study of Linnaeus's botanical classifications on the one hand and a firsthand engagement in anatomy on the other. His polymath ambitions in expounding fields of natural sciences while breaching the epistemic horizons of *Naturphilosophie* entailed, all along, the identification of a common denominator in the broad and various manifestations of nature, a pattern that could grant nature a universal unity. Romantic *Naturphilosophie* envisioned nature as "intrinsically archetypal" and Spinozistically organic,[38] and Goethe cultivated similar epistemic aspirations when

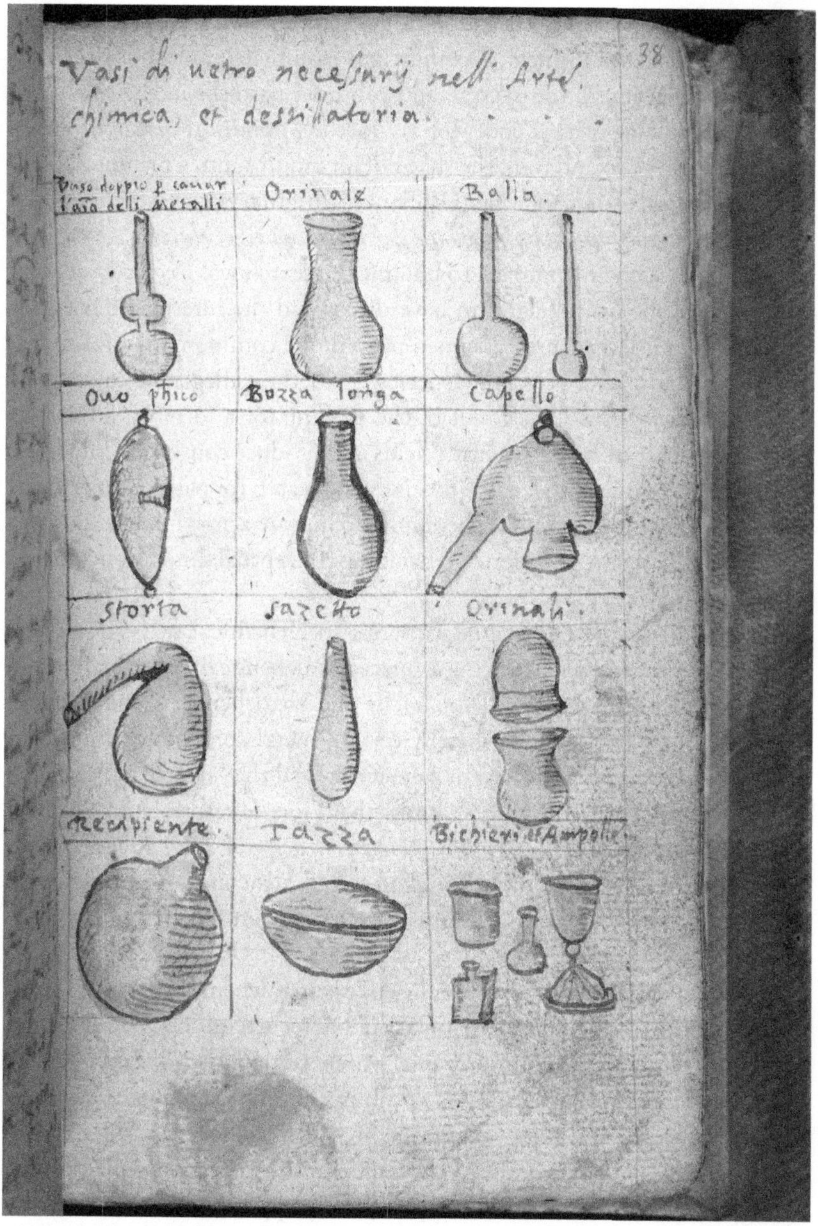

FIGURE 10. Alchemic glass equipment: "Vasi di vetro necessari nell'arte chimica e distillatoria." From Antonio Neri, *Il tesoro del mondo di pietre* (1598–1600). Manuscript. MS Ferguson 67 (GB 0247), f. 38r. University of Glasgow Library, Special Collections.

conceiving his eclectic investigation of the natural world. In particular, he found botanical and Linnean classifications insufficient and thus began his quest for archetypical, transversal, and common patterns in the texture of nature. Exemplary of this endeavor is Goethe's botanical research, which started around 1787 and took the shape of an inquiry into a primordial pattern of botanical expression: the "Urpflanze" (original plant), which resulted in his 1790 *Versuch die Metamorphose der Pflanzen zu erklären*.[39]

Parallel to such an approach to botanical research was his study of anatomy. In 1784, he would claim to have discovered the intermaxillary bone in the human jaw, which he assumed proved one common anatomical feature shared by several vertebrate species.[40] Goethe's allegiance to what he called *Morphologie* was the result of the striving for a form of knowledge that could tackle different scientific fields and produce consistent outcomes. The morphological approach was precisely the search for patterns and forms shared by different objects and phenomena of nature across its variety. It was a knowledge based in intuition and bridging conceptual dimension and empirical experience.[41]

Meanwhile, however, Goethe's interest in alchemical texts didn't disappear: in 1785, he was studying a piece of alchemical experimentation, the *Arbor Martis* (or *Arbor Dianae*) by the iatrochemist Louis Lémery (1677–1743), a phenomenon whereby certain metal compounds transform into plant-like shapes and that for Lémery proved the "sympathy" between minerals and plants.[42] In 1786, we know that Goethe read Johann Valentin Andreae's 1616 book *Die Chymische Hochzeit des Christian Rosencreutz*,[43] which contains a wealth of symbols of mixed alchemical and hermetical origin, the importance of which for the figure of Homunculus will become clear in the next section.

From the vantage point of scientific endeavors, alchemy provided Goethe with a historical record and an exemplary praxis for attempting to understand and reduce nature to its basic elements, which, recombined, would yield the possibility of obtaining substances and transmuting them constantly.[44] At the same time, the symbology employed in alchemical texts as protective secrecy offered Goethe a worldview grounded in the microcosm–macrocosm analogy, hierarchical order, and complementary oppositions conceived as an organic unity of matter and spirit.[45] Together with this, it yielded a figurative language deeply embedded in the concrete experimental activity of understanding nature as a unified system to be decoded—a language that would prove productive on the grounds of his literary work by lending images, structures, and tales. In fact, the very iconographical history of alchemy revolves around ciphered, allegorical images representing humans in acts of intercourse or pregnancy. The genre was established firmly in the Middle Ages and was intended as an effective memory aid. It yielded, however, a coded repertoire that crystallized into typically alchemical images such as that of the hermaphrodite, representing a fundamental mixture of Mercury and Sulfur, also called *Rebis*.[46] While these illustrations represented chemical processes, their iconic power became charged with symbolic possibilities. Before even conceiving of the homunculus, alchemy had already devised an iconography of humans in bottles (fig. 11).

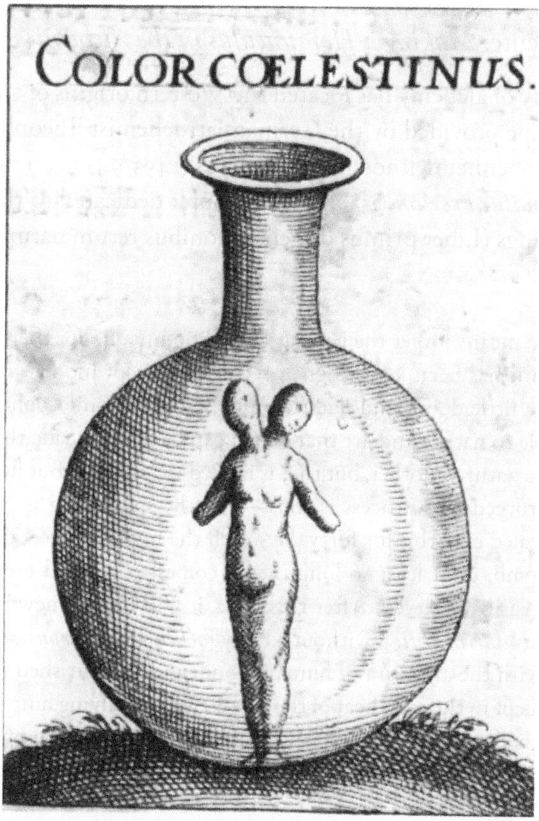

FIGURE 11. Alchemic procedure represented anthropomorphically. "Color cœlestinus." From Johann Daniel Mylius, *Anatomiae auri sive tyrocinium medico-chymicum* (1628), part 5, p. 8. München, Bayerische Staatsbibliothek.

While later also taking a critical distance from certain alchemic traditions as too obscure, Goethe was persistently driven by the range of processes that spread on the alchemical-chemical historical continuum.[47] Be this as it may, alchemy, as a science invested during its Romantic revival with the aspiration to organic totality of Romantic *Naturphilosophie*, provided Goethe with a productive spectrum of operations and interventions into and about nature that problematized the notion of nature in general, as well as of human nature: exactly the set of questions undergirding the tale of Homunculus in *Faust II*. More specifically for our inquiry, alchemy provided Goethe with a key theme—the figure of Homunculus, extracted from alchemy's tradition and repertoire of experiments.

Wunderleute: *Making a Homunculus in the Alchemical Tradition*

The history of alchemy has located the Western origins of the homunculus in the recipe provided by the German iatrochemist Theophrastus Bombastus von Hohenheim, known as Paracelsus (1493/94–1541), specifically in his work *De natura rerum* (1537).[48] In a chapter dedicated to the generation of natural things (Liber primus de generationibus rerum naturalium), Paracelsus writes:

> We must by no means forget the generation of homunculi. For there is something to it, although it has been kept in great secrecy and kept hidden up to now, and there was not a little doubt and question among the old philosophers, whether it even be possible to nature and art that a man can be born outside the female body and [without] a natural mother, but that is indeed possible. . . . But how this should happen and proceed—its process is thus—that the sperm of a man be putrefied by itself in a sealed cucurbit for forty days with the highest degree of putrefaction in a horse's womb, or at least so long that it comes to life and moves itself, and stirs, which is easily observed. After this time, it will look somewhat like a man, but transparent [*durchsichtig*], without a body [*o(h)n(e) ein corpus*]. If, after this, it be fed wisely with the arcanum of human blood, and be nourished for up to forty weeks, and be kept in the even heat of the horse's womb, a living human child grows therefrom, with all its members like another child, which is born of a woman, but much smaller.[49]

The ghastly operations described above should appear quite reasonable once seen from the perspective of the early modern view of life as always beginning with putrefaction.[50] Accordingly, alchemical texts, in particular

De natura rerum, dealt extensively with the process of *palingenesis*, such as the rebirth of birds from their own ashes if mixed with rotting substances and heated in a *venter equinus* (a horse's womb).[51] The belief, which would last long afterward, that the male semen was a miniaturized human needing the passive female womb to mature (the so-called theory of preformation) explains the plausibility of such an imaginary process, devised here through a peculiar form of surrogacy.[52] Nonetheless, Paracelsus defines the possibility of this laid-out procedure of artificial creation a supreme "mystery" (*Geheimnis*; *eine der höchsten und größesten Heimlichkeiten*) and a miracle (*Mirakel*) performed by God.[53] Paracelsus's sources go back not so much to the somehow-similar medieval legends of the Jewish golem created from soil but to the Arabic literature on the creation of artificial animals, in particular to the pseudo-platonic "Book of Laws," also known in the Latin Middle Ages as *Liber Vaccae*, which provides similarly grisly recipes for the fabrication of an artificial human through successive mixtures of organic matter and mineral powders, the aid of some precious stones, and even more crucially, the employment of a cow's womb.[54]

Beyond the bizarre instructions provided, the creature thereby generated was described as an incredibly useful instrument of knowledge and clairvoyance. Indeed, in Paracelsus's text, homunculi are characterized as very special kinds of creatures. According to Paracelsus, the homunculi themselves can yield (though it is not clearly stated through what sort of distinct processes) all kinds of wondrous people (*Wunderleute*), including dwarfs, nymphs, giants, and other creatures of the forest (*sylvestres*).[55] Like all these other creatures, the homunculi can be used as instruments and tools (*Werkzeug und Instrument*) because they know all kinds of secret things unknowable to ordinary humans (*die alle heimlichen und verborgene Dinge wissen, die allen Menschen sonst nicht möglich [sind] zu wissen*).[56] The author then proceeds to explain these exceptional qualities, suggesting the following logical sequence: because they came to life through artifice, their bodies and their flesh are a product of art, hence they cannot learn from anybody, but can only teach. Finally, the author concludes, their powers and actions (*Kräfte und Taten*) make them more comparable to spirits (*Geistern*) than to humans.[57]

The Paracelsian alchemic recipe and description of the homunculus identifies an exclusive set of capabilities that descend from its being artificial—a product of art. In Paracelsus, homunculi are described as little transparent bodies or, better, as transparent entities without a body. Shortly afterward in the text, however, homunculi are described as endowed with flesh and blood, which is at odds with their previously affirmed incorporeality. More than just a contradiction, this ambiguity may hint at the extraordinary and complex nature of the embodiment assigned (by traditional imagination) to homunculi. Homunculi carry the imprint of a human shape (as we will see in a moment, the "mold"), yet they are *Geister* rather than *Menschen.* They do share this nature with other wondrous creatures, and like those (and unlike humans), they have access to secrets undisclosed to human beings. Homunculi, however, are made by, dependent on, and ultimately instrumental to humans. Their exceptionalism, fully codified in the alchemic literature on homunculi, retains an essential validity for the dramatic function of Goethe's version of Homunculus in *Faust II*, where, as seen earlier, this being is instrumental to Mephistopheles for reading Faust's dream and for leading the way to the classical night of Walpurgis.

As far as Goethe's sources are concerned, even if he hadn't read the text by Paracelsus on the generation of homunculi, he would know about the "recipe" and the general speculation about the fabrication of homunculi through other texts, for example the polemical, anti-Paracelsian book from 1666 by Johannes Praetorius, *Anthropodemus Plutonicus*.[58] Furthermore, as I mentioned earlier, in 1786 Goethe had read Lutheran theologian Johann Valentin Andreae's *Chymical Wedding of Christian Rosencreutz* (1616), which might constitute a more probable source for the homunculus of *Faust II*. The allegorical tale of the *Chymical Wedding*, which takes place over seven days, hints at the process of metallic transmutation through language and symbology that are highly characteristic of traditional alchemic imagery. The transmutation of metals is here represented through the image of a wedding, which was the habitual way to encode the alchemic work of mixing metals and letting them react.[59] For example, the text of the *Rosarium Philosophorum*, an illustrated florilegium originally from the thirteenth century, published alongside woodcuts in 1550, aimed explicitly to illustrate the process of obtaining the philosophers' stone (that is, the agent of metallic transmutation) but still employed a highly coded representation of the steps necessary to the task, starting with the disguise of the marriage between Sol (Sun) and Luna (Moon)—that is, the mixture of mercury and sulfur.[60] In the *Chymical Wedding*, the allegorical wedding, followed by the spouses' decapitation—and a subsequent complex series of transformations involving blood, eggs, and ashes of birds—alludes to the alchemic operations of transmutation. In the text, however, this process does not end with the production of the philosophers' stone, as one would expect from such a book; rather, it entails further operations whereby the humid ashes of the bird, previously dampened by the beheaded spouses' blood, are put into molds that, once heated, yield a pair of homunculi.[61] These additional operations have posed an interpretive problem for historians and readers, and it has been suggested that this figure should, in fact, evoke a process of spiritual regeneration.[62] Even though Andrae's source seems to have been Paracelsus himself,[63] there are significant differences in the way the homunculi are generated and how they look in Paracelsus's passage and the *Chymical Wedding*: "We finally opened our little molds, and there were two beautiful, bright and almost transparent little figures, as never man's eye saw before, a male and a female, each of them only four inches long. And what surprised me most was that they were not hard, but white and fleshy, like any other human. But they had no life, so that I actually believe that the image of Venus was also made in such manner."[64] After their removal from the molds, the two radiant and transparent figures, who nevertheless appear as fleshy as actual humans, are not yet alive. After being fed with drops of the bird's blood, they will grow in size, maintaining perfect proportions: "If only painters had been there, they would have been ashamed of their own art in front of these creations."[65] The two figures grow further, until they reach the size of actual humans, with a superlative degree of beauty: "the picture of Venus was nothing, compared to them."[66] The tale continues, ultimately showing the rebirth of the two figures by means of operations involving fire (which is said to transfer a soul into the bodies); eventually, the duo reappear as queen and king of the beginning, though somehow completely transformed.

It seems preposterous, when reading the whole text of the *Chymical Wedding* with the awareness that it is a coded description of alchemical operations, to attribute further symbolic meanings to the tale of the fabrication of these two figures: the account of their birth and transformation is, in fact, perfectly consistent with the secretive disguise that all other alchemical operations have undergone in the text. This text, however, in the long history of its reception, has been read widely beyond its alchemical instructions, feeding a rich interpretive tradition that has looked in it for both enigmas and their solution.[67] The question remains open, whether the homunculi sequence in the *Chymical Wedding* was just a prosecution of the chemical processes illustrated throughout the text or a shift in allegorical register and purpose—a hermeneutic conundrum that exceeds my focus here.[68] What is relevant here is that this text, with which Goethe was well acquainted, provides a description of the appearance and nature of the homunculi that, in part indebted to Paracelsus and in part distinct from it, contributes significantly to the production of the character of Homunculus in *Faust II*.

The homunculi of the *Chymical Wedding* are transparent yet made of flesh; that is, they are figures contiguous to the human. In Paracelsus, as noted already, they were transparent, without a body, yet in flesh and blood similar to humans, though in their essential nature, they were spirits, *Geister*. All these apparently ambiguous and contradictory characterizations point to the direction of a liminal nature of homunculi—a product of artifice and of nature at the same time, generated through the mediation of the art of the alchemist (the *Kunst* of Paracelsus). In the field of alchemy, the boundaries between nature and artifice are typically porous.[69] Homunculi descend from natural processes activated by artificial intervention and are therefore both natural and artificial. This perfectly double descent grants homunculi natural perfection, as well as artificial exquisiteness. This double quality confers them, indeed, an extraordinary appeal. The description of the homunculi in the *Chymical Wedding* insists on the attractive character of this product of art and nature, which defies the beauty of the most accomplished works of art (epitomized by nothing less than Venus's image).

More than just dual, homunculi's nature is liminal and transitional already in the sources that describe them for the first time. Both in Paracelsus and Andreae, though represented in different ways, homunculi are creatures generated at the intersection of nature and art; they are embodied yet spiritual beings; they are superior to humans in their cognitive power yet subordinate to them in the overall design of creation.

The function of Homunculus in *Faust II* is fully consistent with the description in the alchemical sources, insofar as its liminal nature is concerned. This liminality is given an all-round narrative dimension through the actions of the character, mostly informed by his longing for embodiment in flesh but at the same time instrumental to the action of the plot thanks to his artificial perfection.[70] Homunculus's liminality is also the condition for the inception of a strong narrative of transformation and metamorphosis carried out in the drama, and reflecting the broader arc of Goethe's epistemic and aesthetic inquiry in *Faust*, undergirded by the tenets of Romantic *Naturphilosophie*.

Before turning to the structural aspects of Homunculus's plot to foreground the epistemic challenge performed by this peculiar variant of a transparent human, I will turn to further cultural historical elements that are part of the semiotic layers of this figure, which also attends to other discursive dimensions outside of alchemy.

Being in a Bottle

Homunculus is not the first wondrous case of a humanoid or other types of *Wunderleute* to be trapped in a bottle. In fact, this peculiar setup unavoidably evokes fairy-tale scenarios—what I called, at the beginning of this chapter, the folkloric element of Homunculus's tale. To fully reconstruct the genealogy of Goethe's Homunculus, the alchemic lineage I have just traced needs to be integrated by additional elements retrieved from folklore to account for the extensive mythological dimension deployed in *Faust II*.

In 1824 and 1825, Goethe received the volumes of the *1001 Nights* as they were being published in the Breslau edition edited by Max Habicht and as he was starting to work on the second part of *Faust*.[71] It has been convincingly shown that significant sections of *Faust II* are heavily indebted to schemes provided by the orientalist fairy tales, especially with regard to Faust's quest for Helena.[72] Goethe skillfully incorporated a significant structural element from the genre by ensuring, in *Faust II*, the presence of helpful characters throughout the protagonist's quest. In the case of Homunculus, his journey is punctuated by encounters with counselors selected from a wide spectrum of mythopoetic figures: Thales, Nereus, Proteus, Galathea. In his correspondence, Goethe even explicitly invoked the genre of the fairy tale, defining *Faust II* as a fairy tale of interiority: "ein inneres Märchen."[73]

Even if it is not possible to reconstruct with total accuracy Goethe's knowledge of Grimms' edition of fairy tales, it should be noted that the ninety-nineth fairy tale of the brothers' collection (1812–15) is entitled *Der Geist im Glas*—that is, *The Spirit in the Bottle*—and tells the story of a poor boy's fortunate encounter with a spirit trapped in a bottle who ends up bringing him wealth. Fairy tales frequently feature the figure of a genie confined within a bottle and possessing magical powers that hold the potential to bestow wealth and happiness yet also bring ruin on those who seek to possess it (it is the case of the Grimms' tale, where the spirit's first action is a death threat to the boy who has opened the cork).[74] The figure of Homunculus in *Faust II* carries some of the properties of a fairy-tale character insofar as he is himself a helper-figure (as it appears already in the first scene of his appearance). The instrumental nature of Homunculus, already assigned to it by alchemy, blends here with the fairy-tale connotations of the solicitous spirit in the bottle.

One further tale of trapped spirits is that entailed in E. T. A. Hoffmann's *Der goldne Topf* (1814), a fantastic short story staging the mysterious misadventures of the young student Anselmus in Dresden. Cursed by a witch and lured into a magical world, in the ninth "vigil" of the story, Anselmus finds himself as a punishment trapped in a glass bottle, where he will stay for the length of two more segments, inside the alchemic laboratory of Lindhorst. This is the description of his imprisonment in the glass cage:

You are surrounded by brilliant splendour; everything around you appears illuminated and imbued with the hues of a beaming rainbow: all that you see quivers and shimmers and hums in the magic sheen; you swim, devoid of motion and power in a firmly congealed ether which so presses your limbs together that your mind gives orders in vain to your dead body. The mountainous burden lies upon you with more and more weight, and your every breath consumes more and more of the modicum of air which still drifts in the narrow space around you; your pulse throbs wildly; and cut through with anguish, every nerve tenses and trembles in this mortal agony.

Gentle reader, take pity on Anselmus! This unspeakable torture gripped him in his prison of glass, but he knew only too well that even death could not save him, for did he not recover from the profound unconsciousness into which he had been thrown by excessive pain just as the morning sun brightly shone into the room, and didn't his martyrdom begin again? No limb could he move, but his thoughts bounced against the glass, and stunned him with their discordant vibrations. Instead of the words which the spirit once spoke from within him, he now could hear only the muffled din of madness.[75]

Hoffmann effectively describes a numbing and claustrophobic experience in a prison of glass; for in the glass bottle Anselmus is blinded, his breathing constricted, and the sounds he perceives muffled. Being encased in a glass bottle in Hoffmann's tale is a dystopian form of embodiment where sensory perception is hindered to agonizing effect. Anselmus will eventually be freed and returned to the vastness of a magical world competing with a narrower bourgeois cosmos for his dedication.

Goethe had read Hoffmann's story in 1827; in a diary entry from that year, dated May 21, he writes dismissively of it: "Begun to read *The Golden Cup* [sic]. It didn't go down well. I curse the golden Serpentina."[76] That the short story didn't impress Goethe in its aesthetic setup and aspirations doesn't mean, however, that he wasn't aware of, and possibly struck by, the description of Anselmus's glass incarceration. It is also obvious, however, that the conditions of the incarceration of the visionary Anselmus are entirely different from the in vitro existence of Homunculus, who has never known a different embodiment than that of his precariously delicate translucent body encased in the glass bottle and therefore experiences the vial as a cage only insofar as it prevents him from being human. Crucially, in this regard, Homunculus longs not to be released from the glass bottle, per se, but rather—as I will show more extensively in the following sections—to be generated: to *entstehen*.

The folkloric dimension coextensive to Homunculus's staging further emphasizes the instrumental nature of its artificial existence and its exceptionality. Ready to read Faust's dream, to guide him to the classical Walpurgis Night, and to lead the way by means of his internal light, Homunculus will nonetheless choose to forfeit these quasi-magical attributes to pursue his transformation into a human being. The enchanted features of this figure enhance the dramatic novelty of a fairy-tale character trying to elude his destiny, as well as the enormity of his wish to renounce his glass embodiment (the site of his exceptionality) and to become flesh.

German fairy tales host yet another figure of popular folklore, this time preserved in a *Kunstmärchen* by Swabian author Wilhelm Hauff (1802–27), based on the legend of the little man of glass, or "Glasmännlein."[77] In the story entitled *Das kalte Herz* (1827), this diminutive spirit represents the carrier of conscientious and upright action in an early capitalist world where the lure of easily acquired wealth and its dilapidation threaten the whole society, but especially the poor and simple-minded character of Peter Munk, a charcoal burner in the Black Forest who gets mesmerized by the promises of another powerful *Geist* who bargains Peter's heart as a pledge for promised success. The *Glasmännlein*, who like a genie in a bottle is able to fulfill three wishes, appears as a product of the art of glassblowing, back then present in the Schwarzwald region—his features those of a freshly crafted object, flexible as the glass that is being blown when hot.[78] The fable further plays with the idea of a glass container in a different though significant way, suggesting a general symbolic economy of glass as valuable good whose value, however, transcends that of mere wealth. In the story, the giant Michel, the forest's archvillain, collects glass vases containing the hearts of flesh he swaps for hearts of stone, causing his victims to lose their moral compass and pursue ruthless financial gain without scruple. Neatly labeled in their glass vases, the organic relics of a humanity all too ready to renounce itself, they are kept as precious experimental items in the glass vases of the miscreant.[79] The cluelessly ambitious Peter, accepting the bargain, and incurring moral reprobation, will in the end be saved by the *Glasmännlein*, who will miraculously return him his heart of flesh and his humanity. The association, on the one hand, with the material craft of glassblowing and, on the other hand, with an ethical (and clairvoyant) quality corroborates, in the figure of a *Geist* akin to glass, the function of beneficent and valuable spiritual companion capable of fulfilling wishes that are conducive to moral integrity. "Hooray! You are an excellent little glass man, and rightly are you called a treasure keeper, for with you the treasures are truly at home": so exclaims an enthusiastic Peter Munk, as the little spirit made of glass grants him the fulfillment of his wishes.[80] The teensy creature, diminutive also in its name (Männlein/homunculus), reiterates some of the homunculus's crucial features, indicating something like a folktale complex that locates the ethical desirability of an upright existence in the exceptional material quality of glass itself.

"Ensteh auf eigne Hand!" Experimenting with Humanity

After leaving Wagner's laboratory, the motley trio composed of Faust, Mephistopheles, and Homunculus reaches the Greek fields of Pharsalus, where a Walpurgis Nacht in classical historical style is about to unfold. The three äeronauts (*Luftfahrer*), as they are called in the stage directions, decide to split up so that each can fulfill their own desires: Mephistopheles will follow some mythological figure to indulge in gallantries, as well as his obligatory diabolical, cynical commentaries; Faust will ask for directions to find Helena; and Homunculus will seek a way to *Entstehung*—that is, to be born, or generated. When they reconvene, Homunculus explains he has been wandering impatiently, wanting to achieve existence (*entstehn* [sic]) by breaking the glass vial:

> Oh, I keep floating on from place to place
> and, eager to destroy this vial,
> am hoping to *achieve existence* properly;
> but nothing I have seen as yet
> encourages me to become a part of it.[81]

Despite the inevitable sarcasm of Homunculus's observation of humans, his drive remains unchallenged: to achieve existence, *entstehn*.[82] This notion will be insistently repeated from now until the end of Homunculus's plot. By mobilizing an ambivalent semantics of generation, suspended between natural and artificial processes, and a transformative ontology of becoming, *Entstehung* is the guiding concept to understand the actions of this diminutive and extraordinary semihuman existing within a glass bottle.

In this notion also lies the key to deciphering and locating Homunculus along the timeline of glass humans mapped out in this book. The alchemical ambition to generate a sort of human being both more perfect than humans and instrumental to them holds the same kind of questioning function about the power and scope of humanity that the other glass humans also seem to perform. In *Faust II*, by working up and staging (though not up to the actual final transformation into a human being) the onset of humanity, the tale of Homunculus also provides a novel access to the tales of glass humans we have seen so far. More specifically, Homunculus displays the reverse plot: artificially generated as a glass creature of sorts, extraordinary in his material, fleshless embodiment, he profoundly desires to acquire flesh, losing his exceptional existence within the confines of the glass vial. Goethe's Homunculus performs, as it were, a demystification of glass humans by showing the desirability of flesh over the exceptionalism that adheres to possible forms of glass embodiment.

In the narrative scheme staged by Goethe, the wonders of Homunculus are desirable only to the extent that they facilitate Faust's destiny—that is, the destiny of a still deeply, fervently humanist subject oriented toward a cumulative economy of knowledge and, thereby, power. In Goethe's aesthetic and moral notion of subjective agency, Homunculus is bound to find his gift (*Gabe*) ultimately unsatisfactory compared to the possibilities, limited as they are, of humanity.

The dramatic unfolding of Homunculus's adventure toward human embodiment would seem, then, to echo the plea for a self-determined humanist subject also made (in excess, and as an experiment, as it were) by previous cases of humans exchanging flesh for glass. The goal of a full-fledged, autonomous, sovereign subjectivity emerges here for what it is: a precarious liminal space where embodiment is negotiated, away from the earthly burden of flesh that impedes spiritual control but also away from the diminutive, instrumental nature of beings whose enclosure in glass precludes the immersive, organic, and synthetic relation between a subject and its environment—in this case nature itself.

Homunculus, as he further explains to Mephistopheles during the Night of Walpurgis, is looking for two ancient philosophers, who will soon be revealed as Thales and Anaxagoras; from the words they exchanged, they seem to have had access to the essence of real, earthly existence (irdisches Leben):

> I'm on the trail of two philosophers
> who, when I overheard them, were discussing Nature.

I want to stick by them, since they
are bound to know what real existence is.[83]

Homunculus desperately craves an immersive blending with earthly, natural matter into the life cycle. Mephistopheles, with his habitual impertinence, rebukes Homunculus, saying that the help of two philosophers can only turn out to be misleading because, ultimately, being born happens best by one's own hand: "If you want to exist, do so on your own!" (*Willst du entstehn, entsteh auf eigne Hand!*)[84] Mephistopheles's injunction strikes at the heart of Homunculus's ontological drama: the fact that he is an artificial creation of alchemical expertise. Mephistopheles's words distill the ambition of a humanist program directed toward the elevation of the autonomous subject to a biologically subversive, epistemically radical, self-determined subject. Despite the rebuke, Homunculus follows through on his desire, later joining the two Greek philosophers and explaining that he would like to be born: "I'm eager to evolve myself!" (*Mir selbst gelüstet's, zu entstehn!*).[85]

Homunculus's desire to "entstehn" is repeated throughout the following fifth scene, set at the *Rocky Inlets in the Aegean Sea*, where its wish will finally be fulfilled.[86] In this scene, Thales brings Homunculus to the old sea god Nereus, introducing him so:

The advice we beg is urgent for this boy:
How can he best achieve a real existence?[87]

Achieving a real existence is expressed, in the original German, as the desire to be wisely generated, *weislich zu entstehn*. What might a wise form of generation mean? In the context of the drama, it seems to imply, for once, that artificial, alchemical generation, whatever its advantages and wondrous byproducts may be, was unwise, unlike a natural, proper form of generation. Furthermore, the attribution of wisdom to the notion of generation signals that the generation coveted by Homunculus carries an intellectually positive value, again reinforcing the emergence of two radically opposed models of humanity: one (borne by sly Mephistopheles and embraced by Faust) that escalates the stakes of an autonomous, self-serving humanist subject, the other (initially instrumental to Mephistopheles and Faust) represented by Homunculus's attempt at changing the course of his destiny by renouncing epistemic exceptionality and reintegrating into nonautonomous natural existence.

After being redirected by Nereus to Proteus in order to ask "how life's achieved, how one can change his form!" (*wie man enstehn und sich ver-*

wandlen [sic] *kann*),[88] Homunculus meets the metamorphic god, an obstinate shapeshifter, who finally approaches him, having been attracted by his light. At this sight, Proteus exclaims: "What's this which shines so prettily?"; and then reckons: "A dwarf emitting light! A novel sight!"[89] With the designation of little dwarf, *Zwerglein*, Homunculus is assimilated even more clearly to the lineage of Paracelsian *Wunderleute* of alchemical origin.

At this point in the plot, Thales introduces Homunculus's case to Proteus once more, repeating the desire of the tiny creature to *entstehn* and providing a concise description of the liminal state of Homunculus, only "half complete," perfect in intellectual capabilities yet incomplete in what, rather enigmatically, are defined as "tangible properties," and less enigmatically as the wish to obtain a body, literally to be embodied (*verkörperlicht*).

> He wants advice on how he should develop.
> He came, as I have heard him tell,
> into this world quite strangely, only half complete.
> He's well supplied with mental faculties,
> but sorely lacks substantial attributes.
> So far he weighs no more than does his vial,
> but hopes that he may soon obtain a body.[90]

Consistently repeated throughout the scene, the notion of *entstehn* ultimately encompasses not only the meaning of the scene itself and of Homunculus's entire tale but, quite possibly, the meaning of the entire play. As Karl Kerényi wrote in 1941, "the adventure of Homunculus is the mystery of *Entstehen*."[91]

For Kerényi, Homunculus belongs to the wide cosmos of mythology, as a medieval recasting of the "mythologem" of the "Urkind"—the primordial child.[92] Given the thoroughly mythological setting of the Aegean Sea scene, Kerényi's reading confirms the largely mythological quality of the plot of Homunculus.[93] By locating Homunculus within a fixed repertoire of mythical structures, Kerényi was able to identify its basic functions: the hovering between nonbeing and being, the intermediate sexual nature between a male and a hermaphroditic being.[94]

The mythical function of the Urkind is suspended in the liminal space between existence and nonexistence, which—as we have seen—is central to both Paracelsus's and Andreae's alchemical sources. The Urkind's gender fluidity also resonates with alchemical symbologies of homunculi, where the mixture of two substances (such as mercury and sulfur in the *Rosar-*

ium Philosophorum) is said to yield a third substance, which is described (and represented) as both male and female.[95] More to the point: the third function of the Urkind identified by Kerényi is precisely that of *Entstehung*: a permanently interstitial stage of being not-yet-born. To illustrate this stage, Kerényi refers to a fragment of a Hellenistic Theban vase representing a group of figures including a half-born child (Pratolaos) or original human (Ur-mensch). Unlike this figure, however, owning its conception, in Kerényi's interpretation, to Mitos (personification of semen) and Krateia (personification of strength), Homunculus generates itself.[96] In light of this indication, Mephistopheles's advice that one shall better give birth to oneself constitutes further proof that Homunculus is, indeed, a Mephistophelian creature at the same time that he also is a reinstantiation of the mythical *Urkind*, permanently striving for his mythical reintegration into the natural course of biological generation.

The notion of *Entstehung* sustains the entire arc of Homunculus's actions and ties him to the dramatic architecture of the play as a whole, which ultimately seems to hinge on a radical aesthetic rethinking of generation and generative forces. *Entstehung*, in *Faust II*, functions as a structural notion that informs different levels of the dramatic action and its meanings. But if narrowed down to Homunculus's longed-for transition from artificial embodiment in the glass vial to embodiment in flesh, *Entstehung* designates the unique process of embodiment in human form (Homunculus's alchemic generation as such was not, in fact, called *Entstehung*).

Entstehung, however, is not limited to the dramatic final transition of this character; rather, it designates a range of transformative processes that involve the main character, Faust—and, to varying degrees, other figures in the play—on his quest for a new form of life. As far as Homunculus's *Entstehung* goes, it is represented as a complex, multistage process. It is the metamorphic god Proteus who eventually indicates the solution to Homunculus's problem: if he wants to "entstehn," he should start in the sea:

> You must begin out in the open sea.
> That's where you start on a small scale,
> glad to ingest the smallest creatures;
> little by little you'll increase in size
> and put yourself in shape for loftier achievements.[97]

Proteus is suggesting to Homunculus a gradual metamorphic process initiated by first embracing the seawater element. Essential to understanding the notion of *Entstehung* is that it can be carried out only through a process of metamorphosis, or *Verwandlung*. Thales elaborates on this imminent process by telling Homunculus he shall start his life from the beginning of creation (*Schöpfung*), preparing himself for rapid changes and shifts of shape before the human stage.[98] *Entstehen*, for Homunculus, means performing the stages of creation, going through thousands of shapes, before attaining that of the human; only then will his transformation be complete. Proteus warns Homunculus that, once he has climbed on the god's dolphin-back, and has achieved a human incarnation, he should stop his metamorphosis cycle, since becoming human constitutes his end-goal.[99]

This almost technical sequence of instructions implies an underlying notion of metamorphosis as potentially never-ending process; what is being enabled for Homunculus in the sea is a condensed representation of natural history as metamorphosis. That the human being of flesh shall be the *telos* of Homunculus's transformation appears, then, almost as an arbitrary arrest of a potentially perpetual process. In the metamorphic natural history that underpins the scene, humanity appears as a provisional state of perfection, its desirability dependent not on the completion of an evolutionary process but rather on the provisory and arbitrary halt of an enigmatic, perpetually ongoing, process of change. Goethe seems in this scene to locate, or suspend, humanity as a precarious stage on an open-ended course, the destination of which remains untold and subsumed by the mythical Aegean waves.

The ensuing procession of sea gods finally leads to Galathea's appearance on her shell couch. Galathea appears in her splendor: her throne "shines," and she is a "beacon" of light that "gleams bright and clear."[100] As at the beginning of Homunculus's creation in the laboratory, where his luminous appearance cast its light onto the objects of his vision, here his light is cast on the waters in which his final transformation will take place:

> PROTEUS
> Amid these living waters
> your lamp, now bright at last,
> resounds with a glorious tone.[101]

Homunculus's light is seen from afar by the old sea god Nereus. It is then Thales, who describes the last bits of action taking place in the seawaters:

> He'll shatter his vial on her glittering throne—
> there's the flame, there the flash, and already it empties![102]

With these last spectacular pyrotechnics and their acoustic accompaniment (evoking breakage and shattering), Homunculus disappears from the audience's sight and from the play altogether, leaving behind only a trail of light for the Sirens to behold: "What miraculous fire transfigures our waves."[103] The end of Homunculus, accompanied by flames and flashes, mirrors his beginnings; the breaking of the glass, however, is not staged but only announced by Thales in the future tense ("He'll shatter his vial on her glittering throne").[104]

The notion of *Entstehung*—together with those of *Verwandlung* and *Werden*, echoed and repeated in the distinct plots and characters of the tragedy—is a strong key to dramaturgical unity that critics have widely expounded in order to find in the tragedy signs of hermeneutic consistency, mapping what has been described as a microcosmic-macrocosmic correspondence sustaining the entire drama.[105] Most readings of *Faust II*, indeed, attribute the structural cohesion of the drama to a network of symmetries. The most prominent symmetry, emphasized by scholarship that privileges *becoming* as the tragedy's main theme, is that which obtains in Homunculus's desire to be born, in Faust's "rebirth" in the Greek classical scenario of the tragedy, and in Helena's "rebirth" in Faust's quest and imagination.[106]

Faust's desire is properly to reactivate the mythical past, bringing it back to life (and with it, Helena); but through this transformative process, Faust is born again as a new type of "man" in an organic relation to cosmic laws and the memory of classical culture.[107] This reading locates Faust's myth, elaborated by Goethe, within the broader cultural aesthetics of a revival of antiquity that notoriously ignited across Western Europe around the middle of the eighteenth century. From this vantage point, the process of becoming performed by Faust has even been read as a reshuffling of the modern human into a cosmic subject. Parallel to this recasting of Faust is, however, that of Homunculus, defined as the reshuffling of the Christian and Northern element from an exclusively spiritual principle of life into a cosmological animal subject.[108] Following this suggestion, Homunculus's transformation into a human at the chosen point of arrest of an otherwise continuous metamorphic wave, would point toward the coextensive nature of the human and the cosmic, mapping the historical age of humanity onto a cosmological dimension of natural history and thereby rescaling and decentralizing the humanist age.

The symmetries between Faust and Homunculus are further reinforced in the drama by a correspondence between their shared striving (*Streben*), the eager desire that characterizes them both.[109] Other readers have pointed out the contrastive relation between Homunculus and Faust, emphasizing how Homunculus's drive to action, although certainly Faustian, yields an ironic reversal in one who simply wants to *entstehen*.[110] A similar structural symmetry in the play exists between Homunculus and Helena, one sustained (yet again) by a kind of parallel, contrapuntal re-birth.[111] Katharina Mommsen has emphasized the powerful contrast between Faust and Homunculus, not only in their respective trajectories but in their dramatic functions.[112] This specific way of interpreting the symmetry restores a certain autonomy to Homunculus's figure, which otherwise could be all too easily absorbed, in analogical readings, into Faust's own plot. What really puts the figure of Homunculus in an autonomous light, for Mommsen, is the distinct destination of his trajectory compared to Faust's: while Homunculus moves toward a new ontological state, Faust moves only toward a dimension of appearances.[113] From this vantage point, the contrast enacted by these two figures can be viewed as that between a natural and an artificial way of becoming, where Homunculus represents the natural path, while Faust, by rejoicing with Helena, represents the artificial, which only reaches for the truth of art. While the metamorphosis of Faust is through and toward art, the one of Homunculus is different—surely more difficult and, in fact, even impossible outside of his tale.

Ancient myth is undoubtedly celebrated in *Faust II*, but the plot of Homunculus also crucially reveals the actual risky implications of the dream of a return to classic antiquity—which, in order to succeed, would require an engagement not just with art but with nature.[114] But if the reinstantiation of classical antiquity is certainly a central plot of the tragedy—an idea that is definitely in unison with the neoclassical aesthetic concerns that Goethe programmatically embraced—by granting Homunculus a more autonomous space of meaning, the drama's cosmological dimension emerges as its true epistemic expanse. In performing human embodiment within an open-ended line of metamorphic changes, the tale of Homunculus reinscribes the historical and aesthetic concerns of Goethe in the much larger and more inscrutable scale of human history as just one element of cosmic history. This, in turn, opens up space to further read Homunculus as a crucial figure for rethinking the conditions of embodiment and the foundations of humanity within a distinctly humanist horizon.

From Entstehung *to* Entelechie

Entstehung, understood as metamorphosis (or *Verwandlung*), is part of a larger constellation of notions that captures Goethe's investment in philosophical questions and maps onto his contemporary scientific culture, together with notions such as morphology, entelechy, and perfectibility. Their interplay provides the grid on which the plot of Homunculus in *Faust II* can be fully charted and understood. Only one of these notions, however, figures prominently in Goethe's reflections on *Faust II*: that of *Entelechie*. A note jotted by Friedrich Wilhelm Riemer on March 30, 1833, explicitly discloses this connection:

> To my question, what Goethe meant with the Homunculus, Eckermann replied: Goethe wanted to represent with it pure entelechy, the mind, or the spirit, in the way it enters life before any experience; because the spirit of human beings is already exceptionally endowed, and we don't learn all we know, but we have carried it with us all along. For the spirit, the world has begun early on, before all experiences; the spirit has already seen through all things, before it has even made any experience.... Yes, Goethe had a kind of respect, really, for the Homunculus.[115]

In the passage, Eckermann argues that Homunculus represents "pure entelechy" (*reine Entelechie*), the spiritual life of the human being, on the grounds that humans are already endowed with a spiritual destination when they are born, prior to any experience of the world. Homunculus, then, according to the reported exchange, would present a distilled case of *Entelechie* because Homunculus enters the world as spirit (devoid of a proper body), which makes his task well-defined: becoming fully human.

Further exploring this concept through the tale of Homunculus allows us to focus on the stakes of the notion of humanity and on the various epistemic manifestations of humanism that the drama both mobilizes and interrogates and that the glass figure of Homunculus embodies.

Entelechy, a crucial notion in Goethe's work, entered his philosophical horizon early on but was developed explicitly only in his later work.[116] Entelechy, which—according to the classic Aristotelian definition—is the formal principle of inner finality contained in any being, entered Goethe's aesthetic field through the mediation of Leibniz's monadology, as well as Schelling's thought.[117] Entelechy hence becomes bound in Goethe to his notion of morphology, which, as we have seen, guides his understanding of nature as a regulated system of symmetries and correspondences between elements in the microcosmos reflected in those of the macrocosm but also informs his methodology of retrieving the formal patterns of nature, which move from a basic unit to its more complex recombination through a process of metamorphosis.[118]

Entelechy is a structural element of Goethe's epistemology, defining the laws of development that all beings carry within themselves and that drive them to the fulfillment of their morphological destination through a process of metamorphosis. Viewed in this light, one can see how Homunculus, following Eckermann's explanation to Riemer, could embody the movement—or principle—of entelechy "purely." Homunculus's existence, as it is staged in the drama, is guided exclusively by the pure, unflagging form of the striving—*streben*, or *sehnen*—toward its final form: the human.

In the 1817 poetic cycle *Urworte. Orphisch* ("Primal Words. Orphic"), which Goethe later significantly incorporated into his writings on morphology, he depicted the unfolding of human life according to the principle of entelechy, though without explicitly naming it. The first stanza, entitled *Dämon*, provides, in its last verse, a definition of entelechy, after spelling out the inescapable imprinting of human existence effected by cosmic, planetary forces:

DAEMON
As stood the sun to the salute of planets
Upon the day that gave you to the earth,
You grew forthwith, and prospered, in your growing
Heeded the law presiding at your birth.
Sibyls and prophets told it: You must be
None but yourself, from self you cannot flee.
No time there is, no power, can decompose
The minted form that lives and living grows.[119]

The "minted form" (*geprägte Form*) is the shape imprinted on beings at birth and develops (*entwickelt*) over the course of their lives; it is, in fact, the formal principle of inner finality of being—entelechy.[120] Evoking the prophetic power of the horoscope in marking one's destiny at birth, the poem describes a general principle of finalized metamorphic movement, which seems to span the natural, as well as the moral, world of the individual.[121] It is the morphological unity of nature, knowledge, and art that makes it possible for the principle of entelechy to operate across the natural and the moral world of characters, where it acquires the name of *telos*.

In Goethe's first version of the last (fifth) act of the drama, the stage directions referring to the angels' choir defined the immortal part of Faust carried by the angels as "Fausts Entelechie," but in the last version, it will become "Faustens Unsterbliches." That the immortal remains of Faust were called, in that first version, his *Entelechie* pointed to his ultimate spiritual destination. The celestial body of Faust, his still visibly anthropomorphic soul, was his spiritual destination throughout—a maneuver of philosophical subtlety, attaching Entelechy to Christian soteriology.

Goethe provided (again only through Eckermann's mediation) a more mundane explanation of the notion of entelechy, which clarifies its use in the first version of the drama and at the same time trivializes it:

> Every entelechy is a piece of eternity, and the few years during which it is bound to the earthly body does not make it old. If this entelechy is of a trivial kind, it will exercise but little sway during its bodily confinement; on the contrary, the body will predominate, and when this grows old the entelechy will not hold and restrain it. But if the entelechy is of a powerful kind, as is the case with all men of natural genius, then with its animating penetration of the body it will not only act with strengthening and ennobling power upon the organization, but it will also endeavor with its spiritual superiority to confer the privilege of perpetual youth. Thence it comes that in men of superior endowments, even during their old age, we constantly perceive fresh epochs of singular productiveness; they seem constantly to grow young again for a time, and that is what I call a repeated puberty.[122]

In the passage above, entelechy is bound to a self-serving theory of genius, where the exceptional, superior destination operative in a great mind acts as a form of aspirational immortality (by way of perpetual youth).[123] Since the passage states clearly that not all beings possess the same degree of entelechy, as in the inferior kind (*geringer Art*) the body prevails, as opposed to the powerful type (*mächtiger Art*), the Genius. Entelechy, while looking a lot like a mark of "distinction" (as perhaps sensibility had played out in the previous chapter), operates as a differentiated plan, burdened by an evidently classist theory of genius and, with it, a manifest humanist paradigm defining the purpose of the human. The anthroposocial differentiation of human entelechies introduces and justifies an ambivalence that oscillates between the glorification of human embodiment and of spiritual transcendence. It is the same ambiguity that keeps returning as a productive core of meaning for all the glass humans we've seen so far, including Goethe's Homunculus.

Homunculus's plot parallels, complements, and reflects that of Faust; but eventually it follows an opposite trajectory. Because Homunculus's destination is embodiment, *Verkörperlichung*, his path, for the fictional time of the drama, goes in the opposite direction of Faust's, which will reach to his soul's celestial ascension: the two entelechies are not simultaneous, and it would be forceful to just read them symmetrically. Instead, I want to point to Homunculus's entelechy as a foil to Faust's ambitions and designs. By choosing embodiment, Homunculus's supernatural attributes (those belonging to the realm of the spiritual and making him a *Geist*) will disappear once his transformation into a human has begun; his luminous and transparent nature, bound to his glass cage, will be gone, once the vial is broken. His experimental nature shows how precarious and uncertain the path toward the human can be once the humanist myth of Faust is deposed and the aspiration toward epistemic enhancement gives way to a desire for the human to merge with cosmic natural forces.

Becoming Human—Backwards

As a figure from the alchemical tradition, Homunculus's existence depends on the belief in a continuous process of elemental transmutation and the breakdown of nature into its essential units, which could be artificially recombined to return to the same natural compounds. In that tradition, Homunculus bears the stamp of artificiality, on its body and on its features, occupying a threshold between nature and artifice, clearly signaled by its liminal phenomenology of diminutive, hermaphroditic, and spiritual being. It is perhaps due to this ambiguous position that the recipe for the creation of Homunculus remains a marginal episode in the alchemical tradition (somehow liminal to alchemy itself), often charged with an exclusively allegorical function.[124] Unambiguous in the alchemical sources, however, is the description of the possible outcome of the artificial creation of the human, if it ever took place: a little human with a body that is not quite a body and with some exceptional features that would make these creatures useful to their creators, such as the ability to read through the present and to know all the secrets; possibly, too, they would possess an incredibly enjoyable beauty, superior to that of art.

Goethe picked from all these attributions the character of his Homunculus, who, however, reversed the instrumental destination of the alchemical creature into an ardent desire to return to nature, escaping the artificial interstice of his man-made existence. Goethe's Homunculus descends from the alchemical tradition only to relinquish it the moment he chooses to *entstehen*.

At the start of his life in the Laboratorium, Homunculus had told Wagner how curious it was that "what is natural takes almost endless space, while what is not, requires a container."[125] The predicament of Goethe's Homunculus was already fully announced in those two verses: to be natural and thereby lose any kind of boundaries or to be empowered by the very boundaries imposed by artificiality? The predicament is turned by Goethe into Homunculus's entelechy as his *telos*, guided by the aspiration to emancipate himself from the closed space of his artificial existence in the vial—his life in glass. If, on the one hand, the triumphal scenery on the Aegean shores indicates that the return to nature, in the form of the water element, is worth celebrating, we know that, on the other hand, in rejoicing over the natural and, ultimately, human world, Homunculus will also become intellectually obfuscated (*verdüstert*) and limited (*beschränkt*), contained in a different kind of limiting form, the one that pertains to human minds in their flesh bodies.[126] By eventually embracing the limited yet boundless corporeality of a human being against the transparent but instrumental nature of its artificial embodiment in glass, Homunculus makes an exemplary choice in choosing flesh over spirit at the cost of replacing one kind of limitation with another. Homunculus's plot makes the stakes of this choice visible by presenting the entire process of embodiment in glass: the conditions of a glass-encased life and spiritual existence, their wondrous epistemic possibilities in the form of thoroughly transparent knowledge, the forces arrayed against them—fragility, enclosure, and the restless longing for a body of flesh—and, finally, their renunciation and overcoming.

In light of such a paradoxical development, the posthuman qualities of Homunculus are announced and withdrawn at the same time.[127] Through the tale of this *Männlein*, and especially through the liminal conditions of his embodiment, Goethe has doubled, as it were, the experiment in the glass vial. By letting artificial life renounce its exceptionality, the clarity of spiritual enlightenment to be dimmed, the thin boundaries of the glass to be replaced by the thicker barrier of flesh, Goethe has assigned Homunculus a crucial function in displaying how to take up the burden of flesh and forsake *in vitro* existence. The transition will cost Homunculus the epistemic conditions of transparency granted by the glass body—a perfectly unimpeded relation between subject and object, with the added benefit, in his case, of a diffuse light effect or a permanently radiant enlightenment function.

By renouncing epistemic transparency, Homunculus will join the natural world that will ultimately endow him with flesh—that is, a new form of limitation—but also with unlimited striving: "what is natural takes almost endless space," or translating more literally: "To what is natural, the entire world is barely enough." Paradoxically, the limited nature of flesh confers on it an innate, permanent epistemic aspiration toward the infinite, thereby reintegrating the individual into the cosmic order by way of desire.

As a transparent human choosing to vacate its transparency, Homunculus indicates an alternative path for defining the human—one burdened by the weight of flesh but lightened by the insatiable desire to overcome the boundaries of oneself. In this reverse narrative of metamorphosis, whose culmination is a return to human embodiment rather than an elevation to epistemic perfection or unthwarted transcendence of reality, Goethe has distilled epistemic tensions defining Romantic aesthetics at large. Pursuing human embodiment, and learning not to get caught in the ongoing wave of metamorphic mutations, Homunculus seems to do away with an ideal of epistemic transcendence variously cultivated by the Romantic variant of humanist thought and explored from the philosophical horizon of *Naturphilosophie*. By critically engaging with this variation of humanist episteme, Goethe reinscribes Homunculus's human embodiment onto a cosmological vision of nature—one that has always exceeded the human and conferred on humanity an entelechy defined by desire for transcendence. As a dramatic, fully performative experiment, Goethe's Homunculus presents an alternative humanity—one that departs from the ideal of a heightened spiritual existence afforded by a glass body—and moves toward a new transcendence woven into the fabric of natural and cosmological history: opaque, entangled, yet boundless.

FOUR

Twentieth-Century Transparent Bodies

FROM SOCIAL HYGIENE TO NAZI PROPAGANDA

This chapter investigates the fantasy of a transparent human that took on a very concrete and tangible shape in the early decades of the twentieth century in Europe and beyond: a life-size human artifact with a transparent, glass-like body, its anatomical interior meticulously reproduced and entirely visible. While the fantasy of Momus may seem fulfilled by such an object, the interiority it displayed was far from a metaphysics of the soul; it was anatomy, both mechanistically arranged and vitalistically celebrated. Yet this figure—its design, as well as its uses—was charged with a complex cultural symbolics whose layers I will unfold. As I show, this fantasy is entangled with the history of glass humans that preceded it and can be productively read in light of this overarching constellation of kindred figures, which all share, as I have been arguing, the critical function and epistemological ambition of reimagining human embodiment and, with it, the project of the human.

Embedded in the materiality of the transparent human, or *durchsichtiger Mensch*, is the multilayered structure of historical discourses and symbolic projections accumulated until then and seen in their unfolding in the previous chapters. Usually read in the only selective context of the historical circumstances surrounding this anatomical model (be it eugenics, hygiene history, or anatomical history), the transparent human displayed between the Weimar years and the uprise of the National Socialists is, in fact, a composite result of a much longer history explored in this book and, in its historical context, of numerous intertwining discourses. It is precisely this layering—which I will retrace in the course of this chapter—that explains much of its fascinating power, as well as the irresolvable tensions that charge this object, reiterating the speculative and provisory quality of the glass human fantasy as a figure of thought throughout cultural history.

Let us look, then, at the setting where this figure made its first public appearance. In the exquisite Bauhaus-style rooms of the freshly built Deutsches Hygiene-Museum, inaugurated in Dresden on May 16, 1930, tens of thousands headed daily to see the Second International Exposition of Hygiene.[1] On following the signs for the section "Der Mensch," located in a dedicated hall of the building, visitors would reach the entrance of a room called *Der durchsichtige Mensch*: the transparent human. Inside, a see-through statue—a glass-like, cellulose acetate male body containing an actual human skeleton, a web of reconstructed blood vessels and nerves, and the main models of inner organs—stood on a circular pedestal, on an elevated stage, in a semicircular hall. The space was dimly lit, the upright figure the only object on display. In an ongoing show, the lights were turned completely off, and in the dark, the figure began to light up from the inside, starting with the heart. On the back of the stage, lights then illuminated a quote from St. Augustine printed on the wall: "Man wonders over the restless sea, the flowing water, the sight of the sky, and forgets that of all wonders he himself is the most wonderful."[2]

Built in 1926 by Franz Tschackert, a taxidermist at the Hygiene Museum in Dresden, the transparent model was first assembled in a factory that produced marzipan and marmalade. Later, it was moved to the museum building's workshops, where it quickly became a highlight of the collection.[3] By 1935, the model had come to be called the "glass human" (*gläserner Mensch*), as opposed to "transparent human" (*durchsichtiger Mensch*).[4] In the years following its construction, the human anatomical model became so popular that, between 1930 and 1945, the Hygiene Museum had to build at least nine replicas of it, which were sold afterward to other museums or rented out to exhibitions around the world. In 1933, the first female model—a pregnant woman—was designed. In 1935, the first nonpregnant woman was manufactured and sold to the US, where it was put on display in New York in 1936 and then bought by the Chicago Museum of Science and Industry in 1940 (fig. 12).[5] After WWII and the partial destruction of the Hygiene Museum's collections, a new boost in the production of the transparent models, now configured in slightly modified postures, was initiated within the new

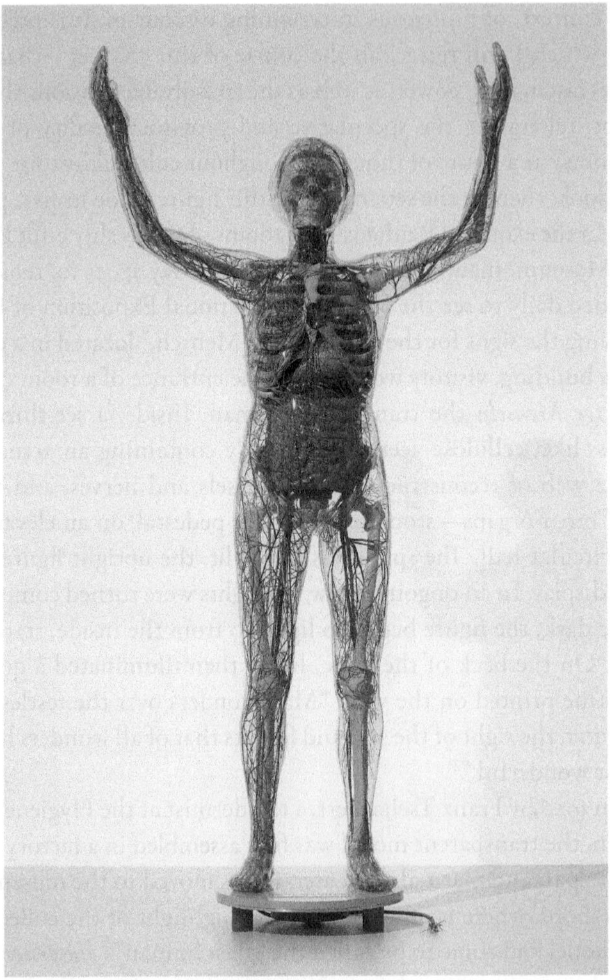

FIGURE 12. *Glass Woman* (*Gläserne Frau*), 1935–36. Photograph by Werner Lieberknecht. Deutsches Hygiene-Museum, Dresden (DHMD 2020/585).

political borders of Eastern Germany (DDR) thanks to the aid of more advanced technologies. Pairs of transparent models—a man and a woman—were delivered to Stalin as a birthday gift and to Mao Zedong.[6] In the 1950s, animal models were also produced by the museum—namely, a few horses and cows—for teaching.[7]

The success of these glass-like figures was extensive, and their presentation was charged with many different, often contradictory, meanings. This chapter follows the production and display of these transparent humans as museum objects during the 1930s in order to read them within the broader context of the culture that produced them and witnessed their popularity. Although these objects continue to inform present-day museal history, mostly as historical artifacts,[8] I will focus here on their emergence and inception in public culture, concentrating specifically on the discursive constellation that attended their production. To reconstruct the cultural space of their display, I look at historical archival material such as museum guides, exhibition booklets, and other photographic documents produced as textual accompaniments for these material objects.[9] By situating these cultural artifacts in their own moment, I show how they were generated at a crucial crossroads in the discourse network of the time. As public didactical models of the human body, these figures are heirs to a long history of anatomic representation, in which the transparent human model represents a distinct moment.

But besides and beyond its place in a figurative anatomical tradition, the transparent human model is imbricated in the emerging discourse of hygiene in the last quarter of the nineteenth century, specifically in Germany.[10] A further, crucial dimension for the discursive location of these figures is the interweaving of the themes of hygiene with the infamous development of German *Rassenhygiene* (a coinage from 1895), or eugenics—that international movement of medical inspiration aimed at pursuing a strategic improvement of the population's genetic pool in the name of social health and productivity.[11] It is through this bleak historical passage that the transparent human model entered the sphere of National Socialist propaganda, which appropriated the apparatus of hygiene exhibitions and incorporated it into its political agenda. This passage is crucial and reflects the way the ideological propaganda of National Socialism operated across strains of thought that were already historically available. As Giorgio Agamben has argued: "Nazism, contrary to a common prejudice, did not limit itself to using and twisting scientific concepts for its own ends. The relationship between National Socialist ideology and the social and biological sciences of the time—in particular, genetics—is more intimate and complex and, at the same time, more disturbing."[12] National Socialist propagandists were able to display the transparent human by transferring to it the new, disquieting projections of a human ideal (the "new human type," in Hitler's infamous formula), conceived as a progressive enhancement of the historical human under the simulacrum of a normative standard.

Through Nazism's historical intervention into the preceding cultural history of the transparent human, however, the broader cultural function of glass bodies is both ratified and coming to an impasse. The humanistic rhetoric of "know yourself," which I addressed in relation to Enlightenment discourses of interiority in chapter 2, is selected in hygienic discourse as the elementary, didactic core of the anatomical exhibit of transparent figures.[13] But the promise of self-knowledge, which was one of complexity and exploration, is replaced, under the rhetoric of eugenic normativity crafted by Nazi propaganda, by a prescription of adherence to a norm.

Looking back at the sequence of glass-human fantasies explored so far—a nonprogressive though narrative unfolding—the early modern episteme of the body conceived as fragile and sacred glass governed by the melancholic complex (chapter 1) gradually gives way to a classical episteme of transparency, in which the subject exposes and interrogates the mechanical ideal of a fully knowable transparent human being (chapter 2). This shift, in turn, prepares the ground for the Romantic fantasy of a life in glass willingly renounced, as the glass body chooses opacity in order to recover flesh and its desires (chapter 3). With the twentieth-century emergence of the transparent anatomical model, however, this critical trajectory comes to an impasse: the critique of the aspiration to exhaustive knowledge of the human is bargained for a fictionally reduced and normatively restrictive standard of humanity, which conceals itself behind a heralded self-knowledge. The transparent human, until then a critical function of humanist discourse, is put to serve the very opposite of its questioning task. The episteme of humanist introspection, however provisory and contradictory, flips into one of acquiescence and compliance.

Yet as an object of material consistence burdened by its own materiality and not only by its symbolic accretions, as well as by false promises and unfulfilled wishes, the transparent human crafted in the 1930s crystallizes into a residual form of material resilience that withstands ideological superimposition. While lending itself to the ideological construct of a "new" human (from hygiene to eugenics to the composite, and often confusing, ideology of a "new human"—*neuer Mensch*),[14] the material imbrication and entangled aesthetic dimension of the "durchsichtiger Mensch" ultimately strikes back, subverting the very normalizing "transparency" it is supposed to symbolize.

To do justice to the entangled symbolic sediment of this figure, this chapter addresses a range of disparate yet closely interconnected topics: anatomical representation, the culture of hygiene and its popular initiatives, the history of eugenics in both German and international contexts, Nazi propaganda, the language and ideas of vitalism, and fascist aesthetics. What the chapter aims to show, however, is not merely the cumulative effect of motives embedded in the glass model but their strategic deployment for ideological ends and their ultimate recalcitrance to full manipulation.

From Anatomy Lesson to Anatomical Show

American readers might have recognized, in the description of the object displayed in Dresden in 1930 and the one portrayed in figure 12, a common fixture of elementary education, which countless schoolchildren have viewed in didactic health centers scattered around the country: a (usually female) mannequin known as TAM (transparent anatomical manikin).[15] Today endowed with a certain outdated (not to say quirky) patina, common to once futuristic objects, these museum items still retain a rudimentary yet effective didactic value. The same educational function was also the prominent feature of the 1930s model, though its educational scope was aimed at an adult, working-class demographic beguiled by the new hygiene exhibitions.

The didactic use of anatomic models is certainly a key function of the historical development of anatomical representations.[16] At the same time, the history of anatomical, didactical representation is imbricated in the aesthetic arrangement and cultural expression of a deep-seated curiosity for the revelation of what is hidden, enveloped, and enclosed. Anatomical models do not thereby exhaust their function in didactic utility but mobilize different forms of spectacle and scopic regimes of desire.

Modern anatomy was designed as a sublimated artistic display of a body of flesh in the illustrations of Andreas Vesalius's *De humani corporis fabrica* (1543). Here, for the first time, human anatomy was drawn with a realistic precision that refrained from abstract schemas. The voluptuous contours of Vesalius's ink bodies, sometimes lifting their own flesh as one would the drapes of a dress, and the anesthetized, normalizing background (insofar as it was familiar to any art viewer of the time), were the result of systematic dissections—the site of anatomical education—while they also displaced the off-putting spectacle of the dissection itself. For Vesalius, this became possible—as the frontispiece of his work makes clear—only after he broke with the long-standing tradition that relegated the doctor to a distant role, directing the barber's hand by voice while the latter performed the "dirty" work of opening the body. Vesalius seized the dissector's tools himself, and his drawings are the carefully crafted, highly mediated products of that first-hand engagement, enacted on the stage of the anatomical theater. In other words, even though his anatomical atlas displayed a sublimated, stylized, and peculiarly sensual version of the human body, the practice of dissection that took place in the anatomic theater—a space of public performance and of unsettling perceptions—lurked behind Vesalius's pictures. His bodies were portrayed in living yet theatrical poses, on the background of a Tuscan landscape—their muscular tone resembling that of the pictorial subjects of contemporaneous Mannerist taste—a compositional ensemble certainly far from the flat, neutralized anatomical figures of atlases perused by today's medical students.[17]

Vesalius's figures may recall, perhaps, the highly controversial plastinated bodies made today by Gunther von Hagens's popular enterprise *Körperwelten* (Body Worlds), an exhibit where perfectly conserved corpses stripped of their skins are immortalized in lavishly active, athletic, or ostentatious poses.[18] Contemporary anatomical spectacles are still at the vanguard of popular exhibitions: since its opening in 1995, *Body Worlds* has attracted millions of visitors curious to watch stripped-bare bodies arranged in action poses—a public driven perhaps by a deeply "necroscopic" desire more than by mere anatomical curiosity, in spite of the scientific frame imposed on the show to legitimize it.[19] Precisely like these exhibits, located on a continuum of what might be called "anatomy art," Vesalius's *Atlas* was not just a didactic tool: it was a spectacle and a huge success.[20]

A functional juxtaposition of the early modern Vesalian body and the transparent model of the Dresden exhibition lays bare the latter's "aseptic" quality, which is reflected in its visual setup, too: not the outcome of a vision from above, as in the dissection theater, but from below, in contemplation.[21] But this diametrically opposed aesthetic rendition of the body was, nonetheless, charged with spectacular implications of a different kind. The massive flow of visitors in Dresden was, in fact, not just attracted by the popular Hygiene exhibition but by the transparent model in particular. It was, indeed, displayed as part of a public show, where its presentation provided a sensory-emotional experience similar to that of viewing a work of art. Advertised solely as a didactic tool, the glass model was certainly more than that; it was a source of visual entertainment. Yet the fascination produced by the transparent model did not stem from the gripping and graphic scenario of a dissection or from the exhibiting of flesh in the corporeal, albeit abstracted, show of von Hagens's *Body Worlds*. In the latter, the public's fascination with the real presence of embalmed flesh may be seen as vicarious to that of a dissecting performance, while filtering out the repulsive sensory experience it implies. Unlike these scenarios, the glass-like model was studiedly aseptic: flesh is altogether absent, and instead of holding its own skin-coat, as in one of the most famous "sculptures" of *Body Worlds* (an open citation from one of Vesalius's drawings), the glass human, in all its hollow transparency, raises its hands to the sky in a praying gesture.

With a gestural connection to a transcendent sphere, the "durchsichtiger Mensch" has abolished its ties to flesh; elevated on a pedestal, it has overcome an earthly, somatic, carnal existence.[22] If the visitors to *Body Worlds* pay a steep ticket price to brush shoulders with death, or at least its organic, plastinated simulacrum, the paying visitors who went to see the "durchsichtiger Mensch" in the 1930s must have aspired to a differently comforting "immunity"—one secured not at the expense of questionably obtained and displayed actual corpses, but by swapping flesh for transparent visibility. While the anatomical show of *Body Worlds* entails a proximity to real dead bodies that seems to sublimate (and thereby exorcise) the death of the visitor, painlessly entertained by the unnatural relocation and reenactment staged by the anonymous corpses, the transparent model effaced the corporal burdens attached to the human body.[23] The anatomical glass model and the corpses of *Body Worlds* respond to different expectations and mobilize different scopic desires. Death was obliterated from the transparent human model not by upstaging it in a highly unrealistic scenario, as may be said to happen in *Body Worlds*, but by presenting a model of the human body as a spectacle of *disembodiment*: an uplifting suspension of the concerns with disease, decay, and death that the public was confronted with in the course of their visit to the hygiene exhibition.

Hygiene

The context of hygiene exhibitions constitutes the first discursive background to historically locate the emergence and success of the spectacle of the transparent human model from Dresden. Toward the end of the nineteenth century, a number of museums dedicated to the propagation of social hygiene practices started to appear in Germany and Europe as the outcome of state social policies directed toward improving living conditions for the large working class. In Germany, the first *Hygiene Ausstellung* took place in Berlin in 1883, preceded by the establishment of a chair for hygiene at the

University of Munich. Figures such as Max von Pettenkofer, Robert Koch, and Rudolf Virchow were the founders, in Germany, of hygiene as a scientifically legitimated and socially tangible discourse articulated around the themes of prophylaxis of diseases and healthy lifestyle, while hearkening back to the French and English developments of hygiene dating from the 1830s and 1840s.[24]

In the exhibitions that rapidly multiplied in the years following the first, in what has been called a "Hygiene-Offensive" promoted by the Bismarck regime, notions such as the need for light, air, and clean water were promulgated alongside the industry's product placements of items supposed to enhance life's hygienic quality, such as Nestlé infant formula or Odol mouthwash.[25] As Swiss doctor Auguste Forel would put it in 1907, when the notion of hygiene was starting to encompass the more problematic notion of social hygiene, in fact, preparing the next steps for the construction of the discourse of "Rassenhygiene": "the task of hygiene is not to cure diseases which are already present, but to guard against all their causes and thus do all we can to prevent them from attacking particular individuals (private hygiene) and the community as a whole (public or social hygiene)."[26] The notion of "racial hygiene" would appear only later but within the same field that had promoted social hygiene policies, and with recourse to the same language of ableism that Forel enlisted in 1907: "The task of hygiene . . . can improve our race so sadly afflicted with abnormalities and bad deformities."[27]

What may have enabled the notion of hygiene to merge with the notion of "racial" improvement is a complex semiotic process unfolding as the association of multiple heterogeneous notions—for example, that of ethics. As Forel laid bare (or, perhaps, even enabled to emerge in the first place), "the conceptions of social hygiene and of *ethics* coincide in an ideal harmony for which we should strive, however many difficulties and conflicts may arise in concrete cases from the defects of our customs, our laws, and our knowledge."[28] Despite the project's ethical overtones, which strike contemporary sensibilities as emphatically disturbing, the underlying project of the vast enterprise of social hygiene, which will engender the infamous racial hygiene discourse, was—unsurprisingly—fundamentally economic: the minimizing of costs deriving from diseases and accidents and the maximizing of productivity within the production system of the Second Industrial Revolution.

The first Hygiene exhibition to reach an international audience took place in Dresden in 1911, funded by Karl August Lingner, the industrial owner of a mouthwash brand. This event was pathbreaking in the public resonance of hygiene: in just six months, more than five million visitors were attracted to the exhibition, and numbers only grew over the following years.[29] One section of these exhibitions, programmed to relocate each time, was a dedicated pavilion called "The human" (*Der Mensch*), with a subsection designated "The transparent human" (*Der durchsichtiger Mensch*), which at the time, however, only displayed histological sections (indeed, almost transparent slices) of human tissues showcased under glass panes. Although the anatomical transparent model was displayed for the first time only years later, at the Second International *Hygiene Ausstellung* in 1930, the rhetorical appeal to transparent knowledge of the human was already explicitly thematized.

Over the course of the 1920s, the success of hygiene exhibitions only grew. In 1926, the city of Düsseldorf hosted the exhibition "GeSoLei," an acronym for *Gesundheitspflege, soziale Fürsorge und Leibesübungen* (healthcare, social care, physical exercise). Here, too, the section "Der Mensch" and its subsection "Der durchsichtige Mensch" were present.[30] In 1927, the Dresden Museum organized a traveling exhibition: "The human in sickness and in health" aimed, according to its exhibition leaflet, at delivering a "complete hygienic enlightenment" (*umfassende hygienische Aufklärung*) as part of a general education (*Belehrung*) about hygienic principles.[31]

The accompanying leaflets and exhibition guides (*Ausstellungsführer*) from these events disclose some of the fundamental themes of hygiene discourse and practices that were current, their iconographical as well as their rhetorical association. This plan of education (*Belehrung*) worked across the field of anatomy and physiology of the human body, where the audience was assumed to lack basic notions, and an ethical, spiritual vocabulary was implemented to lure the public in. An exhibition guide from a further 1928 traveling exhibition from the Dresden Museum, called *Die Gesundheitspflege* (healthcare), reveals at the level of rhetorical arrangement the spiritual surplus of meaning that the organizers of these exhibitions were infusing into the pragmatic and didactic interface of the event: "'This is you,' one is told here—and 'know thyself,' as it was written on the old Greek temple in Delphi, and even if only some of the visitors will be impressed and realize how, without delving into the bodily innermost regions, it is impossible to understand oneself or to live, or to meet the needs of others, the most beautiful task of this 'human being' will be fulfilled."[32]

The task (*Aufgabe*) of the hygiene exhibition was here presented as a mission with ethical, as well aesthetic, overtones. The classical philosophical backdrop of self-knowledge charged the anatomy lesson with humanist connotations: learning about the human body was defined as a reflexive mode of being human, a higher ethical form of life more fully aligned with the body's aesthetic perfection.[33] By following the hygiene measures recommended during the visit, the visitor's reward would have been no less than elevation to the same heights as the transparent model, visually obliterating all the diseases illustrated in other sections of those exhibitions and all the corrupting, sickening, deforming effects of an unhealthy lifestyle. The strategic location of this museal artifact within the hygiene exhibition ratified its function as an object and tool of visual politics.

The glass model, as I have mentioned, appeared at the German traveling exhibition for the first time in 1930, where it was granted a special section. The exhibition guide to the 1931 International Hygiene Exhibition (mostly a repetition of the one from the previous year) provides an overview of the structure of the display and a tangible topography not just of the event but—figuratively speaking—of the discursive interwovenness produced by and reiterated through the exhibition itself.[34] The ground floor of the museum was entirely reserved for the section "Der Mensch," with a subsection for the transparent model; the first floor was occupied—alongside "Hygiene public education" (*Hygienische Volkserziehung*), where a wide range of diseases from tuberculosis to STDs was surveyed—by a section dedicated to nutrition (*Ernährung*) and whole new sections such as "Woman as Wife and Mother" (*Die Frau als Gattin und Mutter*) and "Heredity and Eugenics" (*Vererbung und Eugenik*).

It is precisely the addition of these new sections to the exhibition from 1931 that demonstrates a deeper shift in the conception of hygiene that had taken place over the 1920s, most prominently through the coinage of the eugenic notion of *Rassenhygiene*. The materials accompanying the newer exhibitions record and reproduce the rhetorical layering and association of meanings at work in the museal showcasing of transparent models, thereby revealing their ideological buildup. An example from the description of the section "Der Mensch" in the 1931 guide illustrates the baffling mélange of heterogeneous elements at play in the experience of the exhibition: "The section 'der Mensch' wants to convey understanding for the constitution and *activity* of the human body and thereby awaken *love* for this marvelous *artwork*. The knowledge of its *astonishing organization* must excite a *joyous feeling of the body* as the basis for a sensible and lasting healthcare."[35] The joyful feeling of embodiment, the admiring curiosity for the astonishing human corporeal organization, and love for the physical body understood as an artwork—all these bombastic formulas, embellished through their associative allure, aim at consolidating the pragmatic and normative foundations of the promotion of hygiene.

The logics underlying this use of rhetoric is as subtle as it is commonplace, blending distinct discursive fields through analogical correspondences and thereby producing sweepingly powerful new units of meanings. Another example may further illustrate this strategy. The 1931 guide to the abovementioned exhibition boasts: "According to the dynamic approach of our time, here in the foreground is the active, working, living organism."[36] The three subsequent adjectives—*tätig*, *arbeitend*, and *lebend*—evoke multiple, discrepant elements that are merged into a discursive amalgam. The notion of organism evokes the living unit endowed with intrinsic finality that a new current of Vitalist philosophy was redefining in the first decades of the century, especially through the philosophical work of Hans Driesch. Driesch, who worked as a biologist and conducted embryological experiments on sea urchins, tried to redefine the philosophy of life phenomena by means of a distinct epistemology. Rejecting Mechanism as an inadequate descriptive tool for biological phenomena, he set up a radically alternative vocabulary of life, since "life is a thing of its own kind."[37] The defining notion of the living being was the Aristotelian entelechy (*Entelechie*), the inner finality of life.[38] While technically not a Vitalist (and a foe of the notion of entelechy), Driesch's teacher in Jena, Darwinian biologist Ernst Haeckel, contributed, through his theory of Monism, to the Vitalist vocabulary that would appeal later to National Socialist propaganda. In 1904, Haeckel had published an informative work entitled, very much close to the Berlin exhibition, *Die Lebenswunder* (the wonders of life).[39]

In the rhetorical blend at work in the guide to the 1931 exhibition, the Vitalist finality was then flattened onto the pragmatic finality assigned to a "working" organism—that is, not the Vitalist entelechy but the external finality of productivity forced onto the definition of the human by the economic laws of capitalism. Through this strategic combination of clashing concepts and epistemologies, purposefully arranged to appeal to disparate vocabularies and dispositions, the 1931 *Hygiene Ausstellung* produced a new, bewildering mix of Vitalist undertones with an industrialist-mechanistic representation of the human being also promoted throughout the exhibition. In its spaces, machine-like representations of the body, as in the "human factory" image made popular by Fritz Kahn's illustration *Der Mensch als Industriepalast* (1926), were all over the place, conveying a visual idea of hygiene as the practical maintenance of the body's mechanisms. It was also through the seamless technical rendering of the body that the transparent model illustrated the feasibility of hygienic measures by portraying the body as a machine that had to be preserved through simple, straightforward, routine operations.

The subtle overlapping and association of different meanings is essential to understanding the success and spread of the transparent anatomical model. First, the glass-like, aseptic body allowed the spectacle of anatomy to sublimate into disembodiment and to attract crowds of visitors. On this ground, the inscription of the model within the discourse of hygiene as an industrial offshoot of capitalist productivity was responsible for the staging of the parallel spectacle of the "body-machine," a trusted image of embodiment fantasies (as we saw in chapter 2) and a comforting representation of the body as a mechanism that, with some care, could lastingly be maintained in perfect shape. But this inscription was then combined with a Vitalist vocabulary designed to entice visitors, reconnecting them to a loosely mysterious—perhaps even metaphysical and in any case exceptional—dimension of experience opened by engagement with the anatomical model in the exhibition's space.

Eugenics

The political use of the glass human model doesn't stop at the hygiene exhibitions with their productivity-oriented, ideologically soothing representations of the body as an aseptic construct and an efficient machine. In the German political transition to National Socialist rule in the 1930s, this object gets caught up in the transition, too—a reconfiguration made rhetorically possible by the association of this object and its original hygiene frame with eugenic talking points.

The history of eugenics in the twentieth century is not an exclusively German history; on the contrary, the eugenics movement had international proportions and developed its contents independently from the idiosyncrasies of the German context. A term coined by British polymath Francis Galton in 1883, *eugenics* started off as a movement largely linked to progressive, socialist ideals that were sweeping Europe and North America in the first years of the twentieth century and amounting to reform measures such as clean cities and child welfare.[40] It was in North America that the first proper eugenics policies were established: American sterilization laws—later infamously praised by Hitler as exemplary precedents—were meant to prevent the "unfit" (at the time, psychiatric patients) from reproducing.[41]

In Germany, the constitutive elements of the eugenics movement were first translated at the end of the nineteenth century (in 1895) with the coinage of "Rassenhygiene" (race hygiene), proposed by one of the pioneers of German eugenics, Alfred Ploetz, who was also the head of two main venues for German eugenics. In 1904, Ploetz founded the journal *Archiv für Rassen- und Gesellschafts-Biologie* (Archive for the biology of races and society) and in 1905 the *Gesellschaft für Rassenhygiene* (Society for race hygiene).[42] As Sheila Weiss has extensively shown, the origins of the German movement were, in fact, not immediately anti-Semitic or pro-Aryan; rather, they would become so through the later incorporation of eugenics by the National Socialists.[43]

Ploetz's notion of *Rassenhygiene*, based on the idea of controlling reproduction in order to select an allegedly stronger genetic stock of a population, extended the notion of "hygiene" as a public health issue to encompass the problematic concept of "race."[44] Some of Ploetz's ideas had already been formulated, in Germany, by Wilhelm Schallmeyer, in what is considered the first German eugenics tract: *Über die drohende körperliche Entartung der Kulturmenschheit* (Concerning the Imminent Physical Degeneration of Civilized Humanity) from 1891. Schallmeyer's main idea was to regulate reproduction in order to control genetic heredity, a theme that he developed further in his *Vererbung und Auslese im Lebenslauf der Völker* (Heredity and selection in the life history of peoples) from 1903, where he focused on those measures of eugenics aimed at increasing the "fit" population's fertility rate. As for the identification of the fit group, early German eugenics did not yet hinge on a biological notion of race but rather on a social-cultural version of it. In fact, it was the educated bourgeoisie (the so-called *Bildungsbürgertum*) that was deemed to be the fittest group that should boost its reproduction rate.[45] While Ploetz was infused with a certain enthusiasm for the ethnic Nordic, Schallmeyer never used the expression *Rassenhygiene*, proposing the variant *Rassehygiene* (that is, race in the singular), which was supposed to imply the existence of a single, unified, human race.[46]

The ideological construct of the international eugenics movement, however, with its belief in the selective improvement of the human species, was committed, from its early era, to a notion of "unfit" that laid the foundations for Nazi eugenics. In fact, when Swiss psychiatrist Auguste Forel (under whom Ploetz had studied in Zurich) split the "human race" in two categories—"superior" and "inferior"—in 1903, he was providing German eugenics with the vocabulary of its base eugenics policies (as "inferior," the German *minderwertig*—crystallized).[47] Physical fitness and class went hand in hand for the movement's early ideologists, but Forel made an effort to broaden the social target of fitness: "Modest but healthy, good, reliable, industrious peasants and laborers with a good human understanding are the very best material for a good posterity."[48] This shift seems to register the extension of eugenicist doctrine to larger (and lower) strata of the population, which paralleled its new emphasis on increasing the birth rate, which in Germany had slowed at the beginning of the twentieth century.[49] It is precisely at this historical juncture, when eugenics was at the forefront of social policies trying to popularize its selective principles, that the very first hygiene exhibition in Dresden had taken place in 1911. The format of the exhibition therefore became a major paradigm for theorists of racial hygiene. Eugenics thereby merged inevitably with the discursive topography in which the section on the transparent human (until 1930 limited to histological displays, as we have seen) was situated, further adding to the later object's semiotic complexity.

Through its cross-pollination with eugenics discourse, the transparent human model—hitherto part comforting disembodied simulacrum, part encouraging prototype of mechanistic efficiency—also became a specimen of genetic perfection and racial purity. The peculiar embodiment of this human, aseptically enveloped in a transparent case, which—thanks to its rhetorical presentation—appealed as a wondrous piece of entertainment, would soon sustain the ideological construct of a superior (able and racially sanctioned) human "type" on which Nazi eugenics was predicated.

Before then, in the years of the Weimar Republic, a second generation of eugenicists intensified their racist work, making it available to the emergent National Socialist movement. In the Weimar years, as the conservative and pro-Aryan wing of the movement gravitated toward Munich, the number of institutions dedicated to the spread of eugenics multiplied.[50] In the late stages of the Weimar Republic, the economic crisis triggered a new call for eugenics policies in order to reduce the cost of the welfare state: in 1932, a governmental committee for eugenics drafted a sterilization law designed to establish voluntary sterilization of confirmed genetic carriers of hereditary diseases. This draft was the basis for the Nazis' mandatory sterilization law in 1933, the infamous "Gesetz zur Verhütung erbkranken Nachwuchses" (Law for the prevention of genetically ill offspring), directed toward congenital conditions, including "feeblemindedness," schizophrenia, manic depressive insanity, genetic blindness, and deafness, among others.[51] At that point, eugenics became entirely subsumed under the Aryanist, racist program of the Nazis.

This brief overview of German eugenics shows the sinister continuity between preexisting policies and ideological constructs between the Weimar Republic (and before) and the National Socialist period of power. In the case of eugenics, this continuity is definable both in terms of the Nazi appropriation of its staple elements and vocabulary and in terms of the connivance of scientific eugenicists with the political crimes perpetrated by the Nazis.[52] Whether the discourse of eugenics lent the "form" that allowed the Nazi insertion of the contents of staggering political and social brutality, not entirely entailed in the eugenicist movement, or whether it just expounded a doomed scientific program can be debated.[53] What is beyond doubt is that National Socialism would appropriate not just the contents of a scientific program and its vocabulary—it did not just insert new contents into a preexisting form—but employed a radical rhetorical strategy that, through the ransacking language of propaganda, by means of its cluttered logics, and its specific "pact" with the public, was able to reach a demagogically manipulated public.

Nazi Propaganda

After 1933, all exhibitions that took place in, or were organized by, the Dresden Hygiene Museum—and that included a permanent section for the transparent human model—were attuned to the requirements of the National Socialist German Workers' Party (NSDAP) propaganda committee. The exhibition *Das Wunder des Lebens* (The wonder of life), which took place in Berlin in 1935 and attracted more than six hundred thousand visitors, was designed by the Dresdner Museum in collaboration with the NSDAP.[54] Archival materials documenting this exhibition eloquently illustrate the discursive transition from the hygiene exhibitions of the Weimar Republic to the Nazi-designed exhibitions starting in 1933.

This only apparently seamless transition involving the object "gläserner Mensch"—as the transparent model started to be called around 1935—raises questions about the rhetorical reuse and manipulation of material artifacts and, even more deeply, about the underlying continuity between the discourses that had generated the figure of the glass-like model in the first place (anatomy, hygiene, eugenics) and the National Socialist adoption of the model for propagandistic purposes.

Having sketched the generative discourses for the transparent human model, I now want to concentrate on the specific "insertion" operated by Nazi ideology into this figure and on the strategies of appropriation. This focus will not answer, per se, the question of the deeper continuity between Nazi ideology and its propaganda, and the preceding discursive frames, which would amount to addressing the historiographical problem of ideological continuity between Weimar culture and Nazism—a task that historians of National Socialism have made into a field of research into its own right.[55] My particular focus here is intended to trace rhetorical stratagems that shift an object's meaning by reinscribing it. The type of sources I mobilize to illustrate the strategies of appropriation of the transparent figure are, however, intrinsically problematic in their textual nature. How do we look at a text of propaganda? What analytical tools, what interpretative devices shall we apply? Propaganda is a speech act of its own, a discursive strategy with its own rules and, most of all, its own deceptions. Indeed, the rhetorical status of propaganda itself is highly suspect: in the classic analysis of Adorno and Horkheimer the problems posed by propaganda speech are, in fact, insurmountable, as propaganda mimics (rather than constitutes) discourse, turning any utterance into a lie by precharging it with false contents: "Propaganda manipulates human beings; when it screams freedom it contradicts itself. Mendacity is inseparable from it. . . . It falsifies truth simply by taking it into its mouth."[56] According to the Frankfurt School authors, analyzing a propaganda statement is a paradoxical exercise. As I will shortly demonstrate using a passage from the exhibition guide to the 1935 exhibition *Das Wunder des Lebens*, propaganda content is layered onto—and pressed alongside—preexisting material, allowing the intended message to "pass through"—a process that is intrinsically "intertextual" and yields to an ambiguous relationship to truth. Adorno's and Horkheimer's accusatory plea against the interpretation of propaganda captures much of its problematic textual nature, even if it doesn't exhaustively account for its rhetorical processes.[57] Assuming the Frankfurt critique is an indictment to a textual hermeneutics of propaganda, I will nonetheless try to show some of its rhetorical workings as they occur in the appropriation of the glass-like models by National Socialist ideology.

The Language of Life

By the time of its 1930 debut, the glass human model was already described as posing in the fashion of the fourth century BC "Praying Boy" from Rhodes. From an iconographic perspective, the transparent model, raising his arms in a praying gesture, presents itself as a type, or rather, as a conduit of a *Pathosformel*—what Aby Warburg defines as a gestural pattern charged with a certain expressive content, transmitted across cultural eras but still legible in its original expressive meaning.[58] The iconographical tradition evoked by the gestural presentation of the praying figure assigns it an unequivocal transcendence. While I have already proposed to read this transcendence as an invitation to an experience of disembodiment, enhanced by the contrast with the focus on disease and decay offered by the rest of the hygiene exhibition, the disembodied transcendence signified by the *Pathosformel* of the praying figure is further connotated by the echoes of Vitalism, yielding a modern sacral figure of sorts. In the transparent model, aesthetic elements of the Vitalist discourse and iconological patterns from art converge and reinforce the effect of sacrality.[59]

As we have seen in the section above about hygiene, tracing Vitalist motives from the Neovitalism of Hans Driesch to the Monism of Ernst Haeckel, not only the artifact but the very linguistic accompaniment to the figure reiterates the aesthetic of Vitalism, as we can see in the emphatic vocabulary of "wonder" that plays a leading role in the 1935 exhibition *Das Wunder des Lebens*.[60] Themes of Vitalism articulated through the language of wonder were already present in the exhibition's guide from the 1931 Dresden exhibition. Here, a strikingly dense vocabulary was employed to describe the section on the transparent human:

> As in a *fairy tale*, the eye penetrates in the *hidden depth of the body* and sees through the transparent preparations the structures of the organs so clearly, as never any X-Rays could reveal. Here becomes apparent the beautiful form of the body and of its single parts as a sensual image of the *perfect harmony* that is at work in the *laws of life*. The human being finds within himself *the greatest wonder of the world* if, following the indications of the museum, with a vigilant eye, [he] will look around himself and into himself; this will awaken attention for his body and a deep *sense of responsibility* for the highest good he has to preserve: health.[61]

The "perfect harmony" of the "laws of life" is presented here as the ultimate revelation of that scopic fantasy of looking through the body—a fantasy that X-ray technology had, since 1895, already fulfilled to some extent but that is here surpassed by a stronger desire for (and promise of) transparency and depth. Such fantasy is here morally charged with a "sense of responsibility" toward the body that guides the pursuit of the transparent gaze—a charge that is consistent with the discourse of health policies at the time.[62]

The importance of Vitalist vocabulary is especially crucial to understanding the transition from the 1931 exhibition to the one supervised and organized by Nazi propagandists in 1935. This exhibition, which involved a concrete rearrangement of its sections—giving increasing priority to those devoted to eugenics—also witnessed corresponding changes in the language of its accompanying guides. At the intersection of Nazi propaganda's rhetorical messaging and the preexisting discourses it appropriated, we can observe how the transparent model was reinscribed into an ideological program. In 1935, *Das Wunder des Lebens* brought the transparent model to the German capital for the first time. In the text of the exhibition, it introduced its aims and purposes through an outpouring of life-laden vocabulary, encompassing life as biology and vitality as moral posture:

The great exhibition will work in two directions! For the fellow German, it will relate, in an easy to understand and highly memorable form, the knowledge of the highest organism crafted by Nature to the knowledge of oneself. At the same time, it will reveal the great laws according to which Nature operates, and will deepen our *sense of life*, will strongly reinforce the *will of life*. Walking through the halls of the exhibit, one will fall under the spell of the fact that the solemnity of the exhibition "The Wonder of Life" is accompanied by a strong *sense of joy*, the *merry feeling of the joy of life* of men who, belonging to a great people [*Volk*], are not just the inheritors of a strong past, but carry on a noble future.[63]

A comparative analysis of this text and its earlier 1931 counterpart shows how the rhetoric of hygiene education (*Belehrung*) and filo-humanistic self-knowledge—loaded already with connotations of joy, marvel, and admiration—shifts in the later text to a rhetoric of "joyful" identification with the mythical destiny of the "Volk." This transition occurs, textually, as a mere addition of the new elements to the premasticated contents borrowed from the preceding exhibition—in fact, one of the propogandists' main techniques for appropriating preexisting discourses.[64] In 1931, the notions of "sense of joy" (*Frohgefühl*) and "joy of life" (*Lebensfreude*) were used to characterize, with Vitalistic overtones, the experience of learning about the astonishing "organism" of the body:

The section "der Mensch" wants to convey understanding for the constitution and ability of the human body and thereby awaken love for this *marvelous artwork*. The knowledge of its astonishing organization must excite a *joyous feeling* of the body as the basis for a sensible and lasting healthcare.[65]

In the 1935 text, by contrast, the joyful realization of recognizing in one's own body the meaning of one's existence is linked, with little more than a juxtaposition (in fact, a relative clause), to the ideology of the "Ewiges Volk" (the eternal [German] people), supposedly rooted in a glorious past and destined for an even more glorious future. The Vitalist entelechy previously announced as the marvelous discovery of the perfect human body is here deceivingly made to coincide with the ideological teleology of the Volk. The 1935 Nazi exhibition accomplishes Nazi propaganda's historical appropriation of earlier Vitalist discourse through a sly rhetorical twist and reassociation of images.

The case of the pregnant transparent model—manufactured in Dresden in 1933 and part of a travelling exhibition dedicated to women's health—offers a different example through which to view the strategic appropriation of the glass figures from the hygiene museum into the context of eugenics and racist Nazi propaganda. The female transparent model (fig. 12) did not become part of the German hygiene exhibitions, but a further model—the pregnant woman model—did.[66] Reflecting demographic concerns hinging on gender politics that foregrounded fertility as a primary indicator of female health, this model was annexed to the National Socialist pronatalist propaganda that complemented the antinatalist 1934 law for restricting procreation. Through this framing, the pregnant woman model could become, in fact, the prototype of the "(hereditary) healthy German mother" incorporating the physiological focus on the fertile representation of women into the normative indication of women's commitment to the Reich's eugenicist, pronatalist cause.[67]

On the covers of two further exhibition guides from 1935, designed by Herbert Bayer, we see two highly stylized, similar versions of the transparent human (figs. 13, 14), its fluorescent contours depicted on those of an egg—which was the simplest organic form of entelechy according to Vitalist definitions.[68] Given the special emphasis of the 1935 exhibition on the "gläserner Mensch," and its relevance in the overall exhibition's arrangement, I read in the transparent model's museal objectification during the Nazi era a case of discursive incorporation, on the one hand, and of continuing contentious cultural significance, on the other.

If Nazi propaganda could graft onto this figure its own ideological content, this is partly due to the semiotic frictions that characterize the figure's historical recurrence. The epistemic ideal of a knowable interiority, molded on the scientific and technical gaze into bodily innards, always raised fears

FIGURE 13. Cover of the exhibition guide to *Das Wunder des Lebens* (1935). Deutsches Hygiene-Museum, Dresden (DMHD 2017/1337).

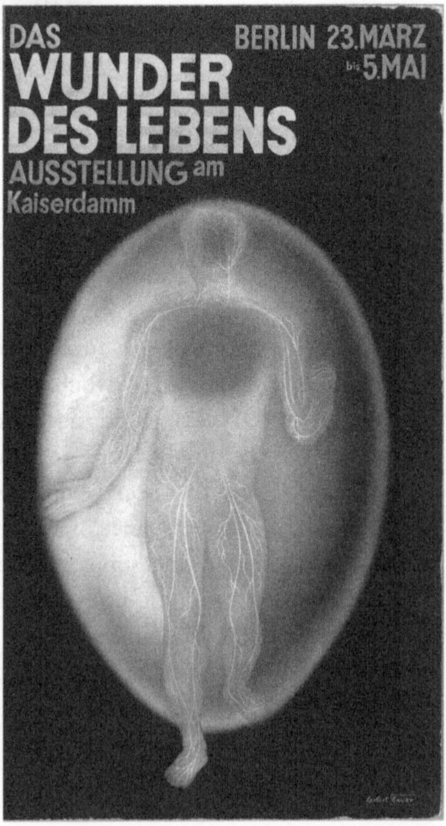

FIGURE 14. Cover of the exhibition guide to *Das Wunder des Lebens* (1935). Deutsches Hygiene-Museum, Dresden (DHMD 2002/1969.1).

of unnatural or forbidden exposure of what is naturally and purposefully encased and protected by flesh, and the liberating thrill of peeking beneath the hood, so to speak, to discover the functions and mechanics of a machine-like being. That is, the transparent figure always (not always at the same time, that is, but alternately and intermittently) occupied the incongruous threshold between a fear of normative exposure as watchable, possibly inexact, sample of a prototypical human ideal and the evasive desire for a fleshless, clarified, and less encumbered existence. The fantasy of the transparent human consistently feeds on these affective sources and their semiotic correlates—sometimes staging their tension, sometimes pursuing its resolution. It is in this regard not exactly surprising that propagandists working on behalf of the National Socialist Party could easily meld the enlightening, socially instructive intentions assigned to the anatomical model with morally and medically normative overtones.

How exactly that happened—that is, how Nazi propagandists were able to incorporate the transparent model as an appropriate symbol for their ideological aims, while integrating it into their own restructuring of hygiene exhibitions under the aegis of heavy eugenics policies—requires, in addition to examining their rhetorical strategies, revisiting the aesthetic arrangement of the transparent figure in the context of hygiene exhibitions. The malleability of its iconological features, medial arrangement, and strategic location within the exhibition's topography displays a network of citations, motives, and patterns that appealed to propagandists, who were able to retrofit them to create a visual semiotic conduit for their ideological program.

Whose Bodies?

When the young Swiss writer Max Frisch paid a visit in 1935 to the Berlin exhibition, he couldn't help noticing—besides the technical show-off of the exhibit, "Spitzenwerk deutscher Technik"—the political implications behind the apparent universalist appeal to "the human" (*der Mensch*).[69] "A healthy human being is a miracle! This is the caption for the first hall, and the exhibits that follow try to make a practical application of our enthusiasm, for all that follows has virtually the same meaning: Only the Nordic human being is healthy and a miracle!"[70]

The 1935 exhibition imbued its promise of universalism with eugenic and nationalist content. In this way, even the iconographic elements of the transparent figure—be it its skinlessness or its posture—assumed further ideological implications.

In the concrete presentation in the context of the exhibition, the glass human model did not stand isolated but was joined by a wide array of bodies didactically and ostentatiously exhibited within the same space. If we look once more at the 1935 exhibition guide to *Das Wunder des Lebens*, we encounter images of muscular bodies stripped of their skin so as to better show the physical wonder of the human body (fig. 15), alongside several stylized and schematic representations of corporal mechanics—both examples of what Hal Foster, following a line of interpretation already pursued by Klaus Theweleit in his seminal *Männerphantasien* (1977), has called the "(proto) fascist obsession with the body as armor."[71] Both figures (the armor and the machine) coalesce under the Marxist analysis of commodification conducted first in Benjamin's critiques and, later, in those of Horkheimer and Adorno. Foster, who deals with the subject of the surrealist representation of mechanized bodies and dolls, addresses the general schema of the "body-

FIGURE 15. Interior page from the exhibition guide to *Das Wunder des Lebens* (1935). Deutsches Hygiene-Museum, Dresden (DMHD 2017/1337).

machine" through Benjamin's critique in the *Passagenwerk*: "Exposure of the mechanistic aspects of the organism is a persistent tendency of the sadist."[72] According to this line of argument, the aesthetic imagery of a body represented as the perfect organization of distinct functions and single, detachable parts—be it in the oversculpting of its muscular features or in the mechanic abstraction of its functions—would reveal a tendency to sadism conceived as both a psychoanalytic dysfunction of identification and as a political attitude toward annihilation: both aspects supposedly staged and unmasked by surrealist art. The protofascist aesthetic of the body is interpreted, in a line that Foster traces back to the Frankfurt School's critique and that he is committed to, both as the outcome of a psychic complex and as a pathological set of social and political tendencies. In fact, Horkheimer and Adorno had formulated this nexus in their *Dialectic of Enlightenment*, where they famously explicitly linked the cult of the physical body to the exterminatory practices of the concentration camps: "Those who extolled the body in Germany, the gymnasts and outdoor sports enthusiasts, always had an intimate affinity to killing, as nature lovers have to hunting. They see the body as a mobile mechanism, with its hinged links, the flesh upholstering the skeleton. They manipulate the body, actuating the limbs as if they were already severed."[73] The passage discusses the "protofascist" aesthetic that Foster also speaks of—that of toned bodies shown in ostentatiously powerful poses. As we have seen, these figures appeared on the same exhibition's guide that hosted the transparent human model and appeared alongside it in the exhibition space itself. In other words, they belonged to the same discursive, visual space that reinscribed the transparent model—continuous with, though not homogeneous to, it.

The glass model was, however, a literally hollow figure with an exposed skeleton and inner organs but significantly no musculature; its body frame was rather unostentatious, not that of a bodybuilder. Perhaps, the body of the transparent human was closer to the image of the "machine-body" schema proper, as, for example, that of the body as factory by Lingner, where the main focus lay in the visibility of the inner mechanism rather than in the chiseled quality of its build.

There is, however, one further strain of bodily iconography, coeval with the transparent model, where the physical ideal of the body and its incorporeal nature converge on a different ground. It is that produced within a yet different cultural-historical aggregate: that of the so-called *Körperkulturbewegung* (physical culture movement), a heterogeneous yet diffuse combination of religious and social utopian motives of renewal, sport, and beauty practices pursuing bodily improvement and thereby the emergence of an improved type of human being.[74] In figures 16 and 17, we see two examples of the artistic aesthetics articulating this cultural rethinking of the body

FIGURE 16. Fidus (Hugo Höppener), *Lichtgebet* (Light Prayer). Study for a stained-glass temple or altar window (1928). Photograph by Marty Kelly. Jack Daulton Collection. Courtesy of Jack Daulton.

FIGURE 17. Fidus (Hugo Höppener), *Luftträger* (1914), cropped. Photograph by Marty Kelly. Jack Daulton Collection. Courtesy of Jack Daulton.

within spiritual reform movements at the turn of the century. The author of the drawings, German painter Hugo Höppener, was a member of the *Lebensreform* movement that took off between Germany and Switzerland in the second half of the nineteenth century.[75] Within this movement, the spiritual and the corporeal elements found a new articulation shaped by meditation practices, bodily routines (for example, vegetarianism), and spiritual beliefs (such as syncretistic and orientalist appropriations within and around the theosophical movement).[76] Iconologically, this blend of beliefs and practices produced the *Pathosformel* of a human being integrated into nature, devoted to health, and at the same time transcending, through physical discipline, its own limited humanity. Simultaneously, as figure 17 shows even more evidently, the cosmic human devoted to light was showcasing a peculiar transparency of the body that hinted at its interior, spiritual (hence, aerial) nature.[77]

Protofascist but also spiritual, armored but also merged with nature, the makeup of the transparent human could appeal to discordant but possibly also overlapping dispositions within its cultural moment.

A visual analysis of the transparent model "gläserner Mensch" is even further complicated by the fact that we are dealing with a multimedia object.[78] At the beginning of this chapter, I described the multimedia experience of a spectacle enabling the viewer to experience identification with a human beyond the fleshy constraints of a human: disembodied, prototypical, and transcending its terrestrial destiny. The special effects that accompanied the presentation of the transparent model, such as the lighting of the inner organs and the acoustic voice-over explanations, inscribed it as a mechanical artifact while also emphasizing the detachment of this model from the community of ordinary living bodies. In the 1930s, surrealist artists were employing automata and mannequins to express "a confusion between the animate and the inanimate, between life and death" through "a doubling of the body; but not just any doubling: its estranging as machine and commodity under capitalism."[79] The transparent model was perhaps bound to a similar duplicity, which, instead of uncanny and critical, was implemented as desirable and assuaging.

Nazi propogandists must have seen in the glass human model an object positively matching their aesthetic values: a perfectly-functioning human machine, a "mobile mechanism"—to echo Horkheimer and Adorno's critical vocabulary—where for once "the flesh upholstering the skeleton" was conveniently removed in order to reveal the underlying mechanism.[80] What these propagandists and ideologists certainly failed to see, or avoided recognizing, was the uncanny doubling effect of the automaton that surrealists, among other "degenerate artists" (as Nazis deemed them), were seeking to rouse with their mechanical figures. The iconographic contents of the glass-like model, in the multimedial context of its actual presentation, represented a continuation of the protofascist aesthetic of the body as much as a departure from it. The transparent model accentuated the aspect of mechanism through its multimedia presentation, up to the point of becoming virtually a mannequin figure—that is, a double, a second-degree representation. This aspect, which the surrealist aesthetic embraced as the "uncanny," remained covered up by the appeal to education and enlightenment that rhetorically framed hygiene exhibitions. Semiotically, however, it lingered as a residual element of representation, an unacknowledged burden that marked this object's instrumental appropriation at the hands of National Socialist propaganda.

Rhetorics of Transparency: "Deutsch sein heißt klar sein"

The infamous speech given by Hitler at the opening of the Great German Art Exhibition in Munich in 1937 proclaimed, in blatantly convoluted and opaque prose, the standard of artistic value for the German Volk in terms of a "law of clarity": "to be German is to be clear."[81] The "clarity" of German art was bizarrely juxtaposed to a corresponding conception of the human body. In fact, in the very same speech, Hitler announced "a new type of man": "better, stronger and more beautiful"—an organism capable of inspiring "a new lust for life, a *sense of joy* in life."[82] Before I proceed with quotes from this speech, it is worth noticing that the vocabulary of novelty applied to the human, that is the notion of a "new human," had solidified in the preceding decades within the aforementioned *Körperkulturbewegung*. Relying on this by then established vocabulary, Hitler's speech was followed by his stereotypical tirade against physical impairments and disabilities, alongside a rant against "primitivist" artists who "suffer from defective vision."[83] The language employed to describe the enthusiastic outcomes of the creation of a "new human type" was highly consistent with that of the 1935 exhibition, *Das Wunder des Lebens*. "Das Frohgefühl der Lebensfreude" (the merry feeling of the joy of life) of the latter announced, rather disquietingly, Hitler's reference to "ein neues Lebensgefühl, eine neue Lebensfreude" (a new feeling for life, a new joy of life). As I have shown, the rhetorical "strategy" of the Nazis had consisted, in the 1935 exhibition, of an "addition"—that is, an insertion of the ideological notion of the "Volk" alongside the Vitalist characterization of self-discovery as an ecstatic and life-affirming experience. This way, the ideology of the Volk had been rhetorically legitimized by means of linguistic juxtaposition, achieving a significant ideological result: the Vitalist entelechy was made to coincide, speciously, with the teleological myth of the Volk. With a similar associative slip, in Hitler's 1937 speech, the ideology of the Volk was once more inserted as a reconnotation of the Vitalistic "Lebensfreude."

In Hitler's speech, furthermore, the "scientific" language of eugenics employed in the exhibition's section on "Rassenhygiene" dissolved into the demagogic rant of demographic cleansing of the "weaker" members of society—which had already become, with the mandatory sterilization law of 1933 (and the secret euthanasia plan), reality. What is most striking about the logically foggy claims of the speech is the key role that "clarity" seems to play in establishing both a "law" for German art and a "law" for a "new type of human." Art and human beings (*Kunst* and *Mensch*) in the ideology of the Volk are presented as homogeneous in their pursuit of clarity—or, better, in their dogmatic presumption of clarity—insofar as they are proclaimed as already clear in their ideological constitution (in fact, were they not yet clear, they would be degenerate [*entartet*]). Meanwhile, eugenics policies pursued by the Nazis in the 1930s, under a scientific alibi provided by a decade-long tradition of eugenics theories, were disclosed as the ultimate extermination program they had become and were assigned the social task of creating a "new human type" through the annihilation of "unfit" humans.

The boisterous clarity that this piece of Nazi propaganda was promoting was the plainness of standardization through normativity, censorship, and physical elimination. Both representational elements such as the "blue meadows" and "green skies" of Expressionist art, and human groups deemed as "undeserving" allegedly shared, in the base logic pursued here, a dangerous lack of clarity, defying the standards established by the ideal athletic body as much as by the "Volk" art promoted by the new Munich Museum.[84] The standard of the "human type" and that of "German art" converged in the muddled formula of *Klarheit* under the sign of normalization.

It is precisely at this juncture that the Nazi use of the clearest of anatomic models—the glass human model from Dresden—becomes apparent. Clarity as an aesthetic notion promoted by Nazi propaganda conveyed (and simultaneously concealed) the ideological contents of violent effacement of ethnic, as much as cultural, complexity. The transparent model, a standard measure of perfectly uniform and plain functioning and compliance to a law of clarity, could become at this point a convenient token of the National Socialist agenda. In 1938, the exhibition "Ewiges Volk" adopted the stylized image of the transparent model for its brochures, thereby iconographically sealing the symbolic political appropriation (fig. 18).

The transparent model, however, as I have shown with reference to Hal Foster's claim about the "body as armor," entailed its own critique in its very material and aesthetic constitution, which evaded the ideological meanings superimposed on it by doubling the human in its *simulacrum*. Today, displayed as historical evidence in hygiene museums, the original glass human model looks very similar to an outdated mannequin: its plastic-

FIGURE 18. Cover of the exhibition guide to *Ewiges Volk* (1938). Deutsches Hygiene-Museum, Dresden (DHMD 2001/1683).

covered skeleton, yellowed by time, exposes the uncanny, deadly face of the ideological construct that was at the basis of its appropriation by Nazi propaganda-designers.[85]

The trajectory of these models' careers after National Socialism opens up further cultural constellations, where Socialist politics in the GDR and world science museums provide new frames in which to read these developments.[86] Changes in design, production, and circulation all attest to the continuing symbolic and discursive inscription of these figures along societal and cultural vectors. But the aging materiality of these only glass-like models—still available to be viewed as spectacles of their own kind—interrupts the crystalline imaginary quality of the intermittent history of glass bodies followed so far.

This rupture in the figure's materiality invites a reconsideration of its semiotic features from the standpoint of embodied viewing. As a way to conclude, I therefore propose an experiment in imaginary embodiment, grounded in the historical clues developed in this chapter and echoing the historical description of the transparent model at the time of its first display in 1930. Let us imagine viewing this model today, at close range. Both the older models and their more recent replicas present an anatomy that is carefully selected, simplified, and readily recognizable—perhaps even reassuring—in its orderly, sober arrangement: an aseptic quality (as argued earlier) that stands in stark contrast to the scene of anatomical dissection. Soothed by this captivating yet not off-putting vision, viewers can now redirect their gaze to the figure's posture—something they might have noticed first on approaching it, were this not merely an exercise in imaginary embodiment. The figure's pose—with arms raised and, at least in the oldest models, a gaze protended upward—shifts the didactic inspection toward a different affective register. The viewer is invited to apprehend a concomitant semiotic dimension: the figure's placement in space and its relation to an elevated, possibly transcendent, object. The figure's postural configuration introduces a phenomenological dimension of existence: what was formerly an object to inspect becomes an imagined subject within a plot either constructed or simply intuited by the viewer. This existential superimposition, however, is not permanent: it is interrupted by the awareness of the figure's objecthood—its transparent casing barely concealing the carefully constructed artificiality of its anatomy, which, in turn, rekindles the viewer's curiosity. While the transparent mannequin evokes a relational, even transcendent, dimension, this gets simultaneously undermined by the constructed artificiality of its exposed interior. The desire that the figure iconologically summons—and that the viewer projects onto it—is tempered by its precise physical presence, redirecting attention toward a matter-of-fact apprehension of its material setup. As unique in its material and discursive arrangement as this artifact is, the semiosis that unfolds in viewing the transparent model seems to confirm, from an embodied perspective, some of the chief epistemological elements of the glass human as a historically recurrent figure: its affective duplicity (fear and delight), its double epistemic ambitions (knowledge of the human as a type and of oneself as an individual), its posthuman status (poised between transcendence and artificiality). This transparent human, viewed today, continues to captivate our curiosity and hold our attention, even as it frustrates our sense of corporeality and our need for a body that, though unburdened by the weight of flesh, may still fulfill the promises of protection and pleasure. In its historical moment, however, these intricacies, which have surely left the stamp of their past urgency on the reflective experience of the viewer as well, were powerfully orchestrated to generate political reactions. The historical viewer was surrounded, immersed and lured into the prescriptions of hygiene, the presumptions of eugenics, and the overt racism of Nazi propaganda through the reminiscences of anatomy, vitalist language, and an ambivalent posthumanist aesthetics that oscillated between new spiritualities, fascism, and racism.

Under the National Socialist juncture, the fantasy of a glass body, which had historically fulfilled a critical function within a humanist episteme, became a simulacrum of humanity's norm built through eugenic selection and biopolitical violence. But an embodied viewing of this figure, imaginary or not, reveals the instability of this ideological construct. The semiotic wavering of this stratified figure keeps feeding its critical function even amid its studied political use; despite the imposing logics of propaganda discourse, the material object resists its symbolic crystallization. The normalizing and destructive fate of the glass body in the history seen in this chapter was not its epilogue.

Coda

This book has traced a cultural history of fantasies of transparent humans—whether as narrative or as material embodiments of glass bodies—analyzed in their cultural moments. These figures simultaneously absorbed and projected conflicting discourses, continually engaging with the implications of humanism and its critique. In so doing, I have opened up the larger semiotics of transparency to the specific circumstances of each figure's emergence, foregrounding the particular discursive and material histories of each glass body over the general conceptual framework of transparency. While this semiotic conceptual framework has shaped the productive area of Critical Transparency Studies, my book intends to contribute to the field by means of a distinct treatment of the glass-body fantasy, which has been lacking so far.[1] By proposing a cultural and literary history of glass bodies, each chapter has honed a particular semiotic constellation that called for both a dedicated timeline and selected conceptual tools.

All the while, as a cultural historical narrative of the fantasy of the transparent human emerged, its exploits have also been interacting with questions and dimensions of transparency at large. Critical Transparency Studies has attended for the most part to the political genealogies of the notion, dwelling on a relatively circumscribed timeline that is relevant for the expounding of the political concept—mostly tracking it back to eighteenth-century philosophy and then following its ventures throughout modernism, where the utopia of transparency peaked, and its subsequent critique in prominent philosophies of the second half of the twentieth century.[2] But this timeline does not coincide with that of the cultural history of transparent humans I have followed. The timeline of this book, first of all, complicates the usual transparency timeline by inserting a historical cluster that predates the usual starting point of transparency discourse by integrating the history of transparent glass bodies of early modernity. Furthermore, the book has dwelled on later elaborations of the transparent human figure that do not coincide with the historical itinerary of the social and political notion of transparency but let a different historical genealogy emerge, one that is not typically part of the record of transparency's timeline—for example, in the case of the Romantic fantasy of transparent homunculi or the Nazi appropriation in the 1930s of the transparent human model. These case studies do not usually fit into the classic genealogy and historical unfolding of the notion of transparency.

The contribution offered by the cultural history of this book is therefore not just one of aligned additions but one of expansion and complication. Not only does the figure of the transparent human disrupt the chronological continuum; it also interrupts the semiotic continuum of the notion that undergirds Critical Transparency Studies. In this book, transparency as a property of the body of the glass human can function in a variety of contradictory ways, depending on the discourses and questions at play. It can function as an empowering property for the knowing and agential subject, a prized condition to showcase interiority, a feared impairment to do the same, an occasion for enacting exceptionality or to renounce it, or an instrument to instruct or surprise, to understand or to conform.

For example, the first chapter examined the transparency of early modern glass bodies as a property transferred on their cognitive qualities: clarity and subtleness, like the properties of the finest Muranese *cristallo*. The second chapter dwelt on quite opposite semiotic properties of transparency: those of exposure in the self-disclosure of the transparent heart of a confessional subject aspiring (or despairing) to become a clear and general object of knowledge. With another variation in the semiotic dimensions of transparency, the third chapter dwelt on the cognitive and epistemic clarity of the Romantic homunculus, a humanoid in glass, but also on its consequent need for an escape from the glass body, thereby reverting the utopia of transparency into its insufficiency. Finally, the last chapter exposed, once more, the transparent human as an object of knowledge and contemplation in the form of a didactic model of health education first and as simulacrum of a normative human type afterward, exposing transparency's instrumentality both to edify and to control.

Each case considered in this book hinges, in fact, on transparency's ambivalent semiosis and amplifies it to critically negotiate, within the productive space of a fantasy, redefinitions of the human. This cultural history doesn't rely on a smooth, parallel correspondence between the genealogy of transparency as a political and social notion (with its aesthetic counterpart) and the history of transparent humans, but rather on a productive, dynamic discrepancy and partial incongruity. Shifting the focus onto the semiotic dimensions enacted by glass bodies affords, however, a far more complex and nuanced exploration of transparency than that played out by the architectural and media-political history of the concept, which is nonetheless intercepted by various instances of the human embodiment in glass.

Since the book's last chapter draws to a close with the anatomical model of the transparent human used by National Socialist propaganda in the 1930s and later delivered to less tragically controversial didactic contexts, a look at any later form of human transparency is warranted: first, to avoid the impression of a teleological trajectory—even one only implicitly suggested—culminating in the Second World War and, second, to bridge the narrative of transparent humans to the present, however occasional or seemingly scarce their appearances may be. The following overview traces the ways different glass bodies have been intermittently generated as fantasies of the human from the twentieth century to the present.

A Man of Glass: A Modernist Fantasy

Modernity's relation to transparency is one of continual topological emergence and critical dismissal. In the last chapter of this book, a transparent human prototype conceived as a didactic instrument for social edification found itself included in a propagandistic program that had to do with the normalization of bodies and the propagandistic fantasy of a human rendered both transparently knowable and commanding of reality. We may now ask how this fantasy relates to the twentieth-century semiotics of transparency explored in the field of Critical Transparency Studies, but to do that, in keeping with the glass body's semiotics, I will briefly look at literary glass bodies that embody representational tendencies practiced and theorized by modernist aesthetics at large.

The fact that modernism seems to hinge on a definite aesthetic ideal of transparent representation is a mainstay of critiques of transparency dominating epistemological thought around the middle of the twentieth century.

As historian of architecture Antony Vidler has argued in relation to architectural history, "the universalism of modernism" was "constructed on the myth of a universal subject."[3] Modernist aesthetics centers on the possibility of transparent representation and celebrates a subject of vision, utterance, and even habitation, whose transparent access to reality grants completeness, reliability, and ethical soundness to the technical and artistic products of its cognition. The epistemic configuration of a desirable, aspirational, even necessary transparency seems, in fact, to regulate modernist aesthetics at large. Within the discourse of literature, it can be traced as much in the theory of modernism as in its practice. In fact, things may be more intricate, and the very aspiration toward transparency in the aesthetic domain may be inseparable from a theory of modernism. A case in point would be the reading of modernism provided by German romance philologist Erich Auerbach in the very middle of the twentieth century. One of the main retroactive postwar interpreters of European literature's representational strategies, and thereby of modernism, too, Auerbach in *Mimesis* (1946) relies on a scheme of representative transparency defining at least the ambitions—if not always the results—of primary modernist narratives that achieve "the reflection of multiple consciousnesses." Such narratives fulfill a promise implicit in the history of what Auerbach understood as "Western" traditions of representation.[4] Auerbach's assumptions about modernist aesthetics relied on a set of epistemological premises that predicated the teleological tendency to an increasing degree of transparency of the world to the subject—what he saw happening in the grand modernist narratives of James Joyce and Virginia Woolf.

The very same modernist understanding of a transparency episteme of literature was accommodated into a figure of a glass body of poetic imagination designed by Wallace Stevens. In his 1940 poem "Asides on the Oboe," Stevens, whose philosophical poetics revolved around the possibility of linguistically retrieving reality, summoned a transparent subject:

> In the end, however naked, tall, there is still
> The impossible possible philosophers' man,
> The man who has had the time to think enough,
> The central man, the human globe, responsive
> As a mirror with a voice, the man of glass,
> Who in a million diamonds sums us up.[5]

This figure was a paradigmatic representative of an alleged rational (Western and male) episteme, a subject of perception and production of meanings whose cognitive activity consisted in mirroring and reproducing reality in the most complete way possible. Stevens provides in his "man of glass" a paradigmatic human prototype designed as "pure responsiveness,"[6] crafted after the same epistemology of modernist transparency at play in Auerbach's examination of literary perspective and representation. At the same time, playing on the semiotic potential of glass—which consistently serves the semantics of transparency and of fragility—Stevens turned this ideal image into its own potential disintegration: the splitting up of a wholesome unity into irreparable fragmentation. In the following lines of his poem, Stevens problematized the centeredness of a subject as a problem of collective knowledge: "He is the transparence of the place in which / He is and in his poems we find peace."[7] The gnoseological comfort of a poet and a subject who can grasp reality transparently, as it is—this had been the aesthetic dream of modernism, a dream that Stevens himself placed on the brink of dissolving to reveal a gnoseological impasse in the face of a historical landscape tinged by "death and war," with the epistemological consequence of an ultimate disbelief in the promises of transparency.

That transparency functions along the semiotic properties of the illocutionary act of promising is shown by important knots of the concept's intellectual history. While modernist aesthetics as captured in Stevens's significantly gendered glass man was both celebrating and mourning the epistemic transparency of the human as a self-shattering ambition, a different modernist subject loomed behind the glass pane of the glass man, enacting his visceral critique. In a short essay containing her existential and literary musings on illness, Virginia Woolf wrote that literature "does its best to maintain that its concern is with the mind; that the body is a sheet of plain glass through which the soul looks straight and clear, and, save for one or two passions such as desire and greed, is null, negligible and non-existent. On the contrary, the very opposite is true."[8]

Woolf provocatively challenged literature on its Cartesian dualist premises, which as we have seen in the course of this book, intercepted the fantasy of the glass human at several points by providing a simulacrum of pure disembodiment through the subtraction of flesh from the body. What Woolf described as a body of glass was, however, a body that, once deprived of flesh, with its epistemic texture—a body perfectly transparent to the mind—would become aesthetically void of existential truth. The body of flesh for Woolf was not only a given but made itself present by never being transparent to its own mechanical functions—as illness reminds us. Woolf's unmasking of the gendered glass man as a fiction of modernist humanism dominating literary discourse and its modernist configuration invoked, from the core of modernist aesthetics, corporeality as the only authentic foundation of subjectivity.

One glass body that echoes and caricatures the wishful fantasies surrounding modernist promises and hopes of transparency is a fictional figure interacting in the multiverse reality of American comics—especially DC comics, a company that began publishing in the 1930s. Here we find in 1947 the figure of Glass Man, a villain who uses a glass mask as a tool to implement a derived superpower to hurt people before being defeated by Batman, who throws him into a furnace.[9] A similar villain is imagined by Marvel comics in 1971, as part of the possible material transformations of Sandman. Fighting with the titular scientist-turned-monster of *The Incredible Hulk*, Sandman's fragility first allows him to have his life spared by the green monster, but as a persistent villain, he later transforms Hulk's partner, Betty, into glass.[10] The villainization of glass figures within American comics operates through the materiality of glass as a source of marvelous and glaring transformations but also through the capricious and yet impending sort of violence on which these narrations hinge, posing some interesting questions about the desirability of those properties granted by glass. In fact, these figures are not transparent, and their glass is materially reduced to a breakable and at the same time sharp and dangerous material. One feature of these figures is their metamorphic setup—something we encountered in the very concrete fantasy (or delusion) or early modern glass humans that developed along the craft and commerce of precious blown glass. Glass is not just the final product of a chemical mixture, the variously thin and transparent, breakable and cutting material of a long, at times even secretive, production process. Glass is also the material of that production process, the soft and incandescent mass plied at the furnace; the persistent, though historically sensitive, fascination with glass is as much with its final properties as with its intermediary material qualities that afford it an extraordinary transformative span. It is of particular interest that a fantasy of transformation such as that of a transparent human (a human that, against all odds of physics, shall become transparent) is also the fantasy of a human made of glass—a material that is intrinsically metamorphic. In fact, a human of glass is a human who *could* readily become other things, as one delusional and inventive man of glass we encountered in chapter 1 knew: throwing himself into a furnace in Murano, he was certain to become a beautiful (and quite useful) glass bottle.

Icicles

As a fantasy invoked by modernist aesthetics and poetics, embodying both its aspirations and failure (the glass man and its splitters), the glass human, in Woolf's boisterous critique, had become an untenable figure of thought. Yet fascination with it persisted in postmodernist literary attempts to pinpoint the epistemic agency of the lyrical subject.

Anne Carson's 1992 narrative in verse "The Glass Essay" dabbles with multiple configurations of glass to locate the cognitive and emotional subject of experience and poetry amid sentimental turbulence and heartache. The autobiographical narrative follows glass reflections and surfaces as one would follow the promise of seeing through others and, thereby, oneself: but prosaically, the encounter with the materiality of glass evades such epistemic illuminations, as in this passage, which echoes closely Mandelstam's passage from *Tristia* I mentioned in my introduction:

> A great icicle formed on the railing of my balcony
> so I drew up close to the window and tried peering through the icicle
> hoping to trick myself into some interior vision,
>
> but all I saw
> was the man and woman in the room across the street
> making their bed and laughing.[11]

The promise of cognitive enhancement attainable through the medium of glass dissolves here into its hollowness. The glass medium is literally empty, only screening the view of a background, not that of an interiority.[12] The modernist scopic fantasy of transparency hence exposes its shortcomings as "a soul trapped in glass,"[13] prompting transparency's dismissal. It is no longer the transparent vision of the other or the encasing of one's soul (according to the vantage point from which glass is being seen),[14] the one informing the lyrical vision of the self and the other, but the radical fantasy of the "nude," of a flesh body stripped naked, to the bone.[15] The image, while reactivating the long-standing topos of "the naked truth,"[16] ties back to the scopic fantasy of a transparency without mediation that, formulated by Rousseau at the start of his *Confessions* as "intus et in cute," coincided with the anatomical gaze sustaining the fantasy of the glass human in several of its materializations analyzed in this book.

What the literary glass bodies of modernist and postmodernist poetics display is indeed a close correspondence to the historical narrative of transparency's conceptual exploits. As Stefanos Geroulanos has extensively shown in his research on philosophical transparency, what follows the modernist ambivalent triumph of transparency (both chased and deemed insufficient) is a general tendency toward the demise of transparency's dream.[17] Foucault's seminal analysis of the panopticon, in particular, established a genealogy of modernity in which transparency acted as modernity's "master code."[18] Aligned with this critical genealogy best grasped by Foucault, the cultural ferment in postwar France generated an extensive deconstruction of transparency, indicting transparency as a normative ideology with the potential of trumping secretive, discordant, and recalcitrant recesses of interiority in the name of virtuous and flat exposure, emerging as the ideal tool for tyrannical and normative regimes.[19]

The critique of transparency that peaked in France in the second half of the twentieth century (especially between 1945 and 1975) erupted as a choral takedown of transparency's chimeras from a multitude of philosophical angles and was pursued by an array of different authors from Sartre to Merleau-Ponty, Althusser, Foucault, Derrida, and Lyotard.[20] Veering from a more closely philosophical and epistemological critique to a properly social and political one around 1968, transparency's unattainability had been decreed: "The world was not transparent, because it was complex, layered, structured, filled with heterogeneity."[21]

Following this narrative, the desirability of a glass body seems to be out of the question, even in the interrogative and experimental form of a fantasy. The glass body of the transparent modernist episteme, occasionally crystallized into a super powerful villain figure, has been seemingly slowly melting back into a liquid form, or, to stay with Carson's vocabulary, the glass-like quality of the icicle simulacrum has revealed its aqueous nature.

The validity of this narrative and its relative historical analysis notwithstanding, the transparent human fantasies described in this book have allowed me to circumvent but also complicate this line, showing how, when epitomized in concrete and variously material figures, the narrative of a transparent figure has also always provided its own foil, its own deconstruction. For example, claiming with Foucault a transparency of the Classical

Age (what in France's intellectual history is located between 1650 and 1800) would become problematic from the standpoint of the very narratives of transparent humans produced by this age, as we have seen exemplarily in the foiling figures of Rousseau's "transparent heart" and Sterne's extraterrestrial glass beings in chapter 2.

It is also true that transparency's history is not a homogeneous one but, rather, is refracted in multiple histories caught in their own chronotopes. In this sense, the French deconstruction of transparency represents a very specific, exceptionally lively, cultural constellation that is not necessarily matched by other cultural contexts, where the supposedly dangerous implications of transparency may sometimes undergo a more forgiving gaze, mostly channeled into a political rhetorics of transparency's necessity and benefits.[22]

It is, in fact, this broader wave of renewed trust in the possibilities of transparency that became increasingly prominent in the 1980s, thanks to the combined action of a prevailing rhetoric of freedom of speech and information and its institutional counterpart.[23] Its effects also partly shaped the glass bodies that emerged as fantasies of digital and medical space.

Glass Patients

Today, the rhetorical domain of transparency as a positive notion to qualify politics and morals may seem uncontested; on the level of governmental policies, its regulative value is taken for granted—transparency a regulative standard of governmental accountability.[24] Political discourse appeals to this notion while also massively manipulating its meaning, producing very ambivalent, if not contradictory, results.[25] Extensive recourse to the semiotics of transparency has not neutralized its contentious meaning. The conundrum can be summed up as the problem of transparency within liberal democracies: their appeal to its value, and the conflict it generates with other mainstays of liberal politics, such as privacy, national security, or internal deliberation.[26]

Transparency's meanings and implications thus permanently seesaw between the semiotics of openness and disclosure claimed by publicity politics and those of violation of secrecy and violent intrusion claimed by critiques of transparency throughout its philosophical and political history. This oscillation is politically sensitive because it is rhetorically unstable, hinging on the uses and the semiotic properties of the media to which it is every time assigned.

In the present age of pervasive digitalization, the meanings of transparency and its values in relation to digital communication are teetering. Surveillance technologies have enhanced the overall transparency index of social life thanks to a ubiquitous digital tracking. Digital transparency can be seen as the enabling of a tracking system of consumer preferences and impulses, masking as transparent accessibility the ideology of information at the service of multiple economic profits.[27] The imperatively positive notion of transparency turns quickly, once again, into its simulacrum, an all-encompassing deception where subjection to permanent tracking at the service of late capitalist economies is sold for transparency of access.[28]

The digital proliferation of traces that profile habits and enable algorithms predicting commercial behaviors based on these profiles may seem to bring the regulative notion of transparency to a breaking point. At the same time, virtual existences unfolding online, and especially on social media, seem to reverse the ambition to transparent consumer profiles into the attempt at masking, filtering, and modifying public identity through the convenience of AI technologies that make it increasingly difficult to discriminate between truth and fiction, and where this distinction, in fact, may not count anymore. What would it mean, for transparency, if the boundary between virtual and real were not relevant anymore to define crisply what things are, who one person is, what identity means? The problem is not new, as we have seen in some of the junctures explored in this book, but the extent to which it is maximized by available technologies and digitalization makes it urgent in new ways. In this perhaps dystopian turn, transparency is implemented as a delusional dispositive to whitewash sophisticated technologies of control over consumers, finances, and the economy of "free" time.[29]

But this book has not been about the exploits and destiny of an aerial concept contended between semiotic clusters and media applications. To state it once more, the object of the story told in these pages—the fantasy, narrative, or material of a glass body—feeds from this conceptual semiotics but is concretely more circumscribed, articulated into the shape of a body, materially cast into a glass body. This fantasy follows its own genealogy, which intercepts that of the abstract concept but does not ultimately map onto it. Indeed, when tracing the contemporary ventures not of the concept of transparency but of the figure of the transparent human, one does encounter again and again a fantasy that, in fact, complicates the conceptual outline of transparency even more than its conceptual, inherently oscillating attributions.

One major contemporary occurrence of the transparent glass body is, as it turns out, a figure of speech at the service of health polices discourse. While figures of speech have intermittently appeared along the storyline of this book, it was often in combination with fictional figures (for example, the transparent heart as the metonymical figure embodied by the eighteenth-century confessional author). Instead, the figure of speech produced by contemporary discourse that I am here referring to is one of public parlance, with a more pronounced frequency in journalistic speech. Already mentioned in my introduction, the "glass patient" has become a notion of health data disclosure in German contemporary culture, crystallizing phenomena and tendencies that are spread everywhere in technological states. In 2015, the weekly news magazine *Der Spiegel* dedicated one of its covers to what was seemingly a sci-fi figure, thanks to the graphics accompanying the article.[30] What in 2015 looked like a new frontier is, ten-plus years after, institutional reality. In January 2025, in Germany, all insured persons received a digital identity (so-called ePA, *elektronische Patientenakte* [electronic patient record]) disclosing all their health data made available in the network of health care. As is to be expected, the newest legal form taken up by this long-standing process has fed robust discussions, from legal scholars to doctors to patients, about the (dis)advantages of exposing such information to the risks of digital networks. Especially when put in relation to the possible commercial uses of sensitive health information, tracking health and conjuring economic exploitation become one and the same agenda (as anybody with access to social media in the age of algorithmic tracking and predictions knows all too well). Such a glass body, as a figure of bureaucratic speech for digital identity, advances the maximalist promises of transparency as access while also rekindling its darker fears of surveillance.

But it is a different application of medical discourse, which has been present in many different forms in this book, that maintains a primary space for the inscription of the contemporary transparent human today, aside from data records. As we have seen, medicine as a field of knowledge stewards the pursuit of inspection of the body in its deepest physical layers and recesses. Medical interventions across the epochs traversed, be it those of anatomy, physiology, or pathology, carried out acts of looking, cutting, investigating, magnifying, modeling, or reproducing the interior space of the body. In the sector of health care, the technologies of medical imaging have fulfilled the promise of a transparent access to the body that belonged to medicine all along—openly so from Vesalius's anatomy through mechanistic models of the body, to the models of the Hygiene Museum in Dresden, to X-ray technology and beyond.[31] While this promise, and its underlying desire, runs in part along a technological and epistemic continuum, the rupturing effect of the introduction of some of these imaging technologies cannot be underestimated or simply resolved into a teleological narrative of progress. To provide an evident and famous example, the discovery of X-rays by German physicist Wilhelm Röntgen in 1895 represented also a break in the visual and symbolical expectations of accessing interiority, reshuffling the topography of interiority altogether.[32] The possibilities unleashed by X-ray technology—at a time when its fatal side effects were not yet clear and its functioning was still primordial—engendered fantasies not just of complete interior visibility but also of disappearance. When H. G. Wells wrote *The Invisible Man* (1897), he was toying with possible outcomes and paroxysmal consequences of the technology that was making the bodily case of humans invisible.[33] By producing a completely transparent, invisible human, Wells was chasing an idea of complete transparency as disappearance and visual camouflage onto the background—a type of transparency that this book has intentionally excluded from its narrative, as explained in its introduction. The transparency associated to the glass figures of this book is, instead, one of heightened visibility, centering on the optical possibilities embodied by such fantasies.

Toward the end of the twentieth century, a very different transparent human than the invisible one was produced by developments in medical imaging. In 1994, one of the main imaging enterprises of contemporary medicine was made accessible to everyone: The Visible Human Project scanned through MRTs and CTs two human cadavers, their bodies sliced by a device called cryomacrotome,[34] producing eighteen thousand images and then making the dataset publicly available.[35] While the imaging ambitions of this project align with those of medical imaging at large, the project represents an unprecedented, "new form of corporeal transparency"[36] based on accumulation and ultra-analytic imaging. The depth of data and analytical reach shifted the visual focus from a human with an inside body to watch, know, and map to that of an inside body made of innumerable layers and whose outer carrier (the body of the human being) was entirely blurred out of focus (even if still implied in the mental map of the expert viewers, the actual users of the database, able to locate the visual fragment within the invisible but implicit body-case). The accomplishment of what imaging technologies in medicine had always strived for, peaked here into a hyperdetailed vision that ended up eliding, or literally excluding from the frame, the human shape, resulting in a visual paradox of sorts. But the human dissected and archived there—the pinnacle of transparency of the human for the aims and purposes of medicine up to that point—evades the morphological contours of this book's narrative. The human of the Visible Human Project has vacated its premises, not just on the basis of its postmortem corporeality but on the very visual devices appointed to its visualization.[37]

The transparent human fantasies of recent decades would seem to differently dissolve into a figure of speech both in the notion of digitalization of health data and in the archive of medical imaging of the Visible Human Project. In both cases, the contours of the figure acting as a human, endowed with a transparent, perhaps even breakable, body, have dissipated: a counterintuitive dematerialization appears to have taken place in the medial expression of the fantasy of a glass body—a fantasy that has otherwise been consistently material throughout this book.

Yet Momus's vivid fantasy and outlined promise to swap flesh for glass to implement a better way to be human, as preposterous and self-defeating as it typically turned out to be in its fictional enactment, keeps on feeding a number of heterogeneous medial occurrences of glass bodies. Sci-fi imagery offers accessible ways to picture and shape transparency of bodies, often within the threatening plots of dystopian future narratives serving the mesmerizing effects of spectacle that accompanied the reembodiment in glass from the start. In a semiotic twist that speaks to the concerns of our present, transparent bodies are often those of AI subjects crafted to simulate humans. In the 2015 sci-fi thriller *Ex Machina*, by Alex Garland, a robot perfectly resembling a human is exposed and undressed, in the transparent casing of her electronic circuits, clearly signaling, through the wrong bodily case, the human/nonhuman divide intended to ultimately soothe the epistemic uncertainty of the human protagonists, as well as of the spectators confronted with an all-too-human robot and the well-known "uncanny valley" effect.

A plethora of recent titles ranging from thrillers to psychology divulgation have made use of the glass human (man of glass, glass humans, the glass man, and so on) to evoke with metaphorical resolve scenarios of fragility or of menacing eccentricity. As a figure of speech, left to the evocative power of its semantics, the glass human retains quite a successful existence as a catchphrase, as in Boccaccio's expression "uomo di vetro" (chapter 1), as an all-explainer for notions that, spelled out, would probably be less headline-grabbing: fragility, exposure, vulnerability, precariousness, paranoia. In *Child of Glass* (2019), an illustrated book for children by illustrator Beatrice

Alemagna, a particular restaging of Momus's promises under the sign of a deep-seated fear of exposure and vulnerability brings up an interesting interlacing of different layers of meanings attained by different historical dimensions of the figure. A girl's thoughts are visible through her forehead, which increasingly becomes a problem as the girl grows up, until she starts cracking on the surface of her body, as if she were made completely of glass, before she decides to embrace her glassy nature with new confidence. That the book is beautifully crafted to introduce some transparent papers at key points of the girl's story is an interesting attempt at rounding up on material concreteness the embodied fantasy of being, possibly, really and truly (as Vidriera in chapter 1 would have put it) made of glass.[38]

Veering into the realm of children's literature, the figure of a glass body may seem to regain for a moment the variously material textures of fantasy whose history has been told in this book—a texture partly unwoven by the tendency of the figure to crystallize in tropes or to migrate into human forms that have vacated their humanness—but it also reminds us of the persistent presence of a figure that seems to function as a *topos*. And as such a rhetorical construct, every subsequent instantiation of a glass body carries the traces of other transparent humans, not only citing but also taking a stance on that tradition. The glass human can then become an erudite quotation, a fragment of cultural history reinserted in the history of the present (or of the future, in the case of sci-fi and dystopic narratives). It is the case of one of postmodernism's seminal novels, David Foster Wallace's *Infinite Jest* (1996), where amid an encyclopedically dystopic world, the transparent human of early modern tradition is quoted, in a footnote, among an incongruously overflowing fictional list of filmic productions, as a fragment of an exorbitant, excessive tradition: "The Man Who Began to Suspect He Was Made of Glass. Year of the Whopper. Latrodectus Mactans Productions. Cosgrove Watt, Gerhardt Schtitt; 35 mm.; 21 minutes; black and white; sound. A man undergoing intensive psychotherapy discovers that he is brittle, hollow, and transparent to others, and becomes either transcendentally enlightened or schizophrenic."[39]

A book of cultural history carries an inevitable cautionary stamp to imagining or predicting the future of cultural objects, yet it seems at least valuable, if not desirable, to keep thinking of transparent humans as our possible counterparts or cautionary tales. Whether this fantasy will open up discussions about the boundaries of the human—its material constitution, its volatile and constructed, perhaps even replaceable existence—is something that only future interactions with this figure—its hermeneutic history, semiotic structures, and, most of all, its radically critical potential—will offer. As long as this interaction will take place, the history told in this book will be just a point of departure.

Notes

Introduction

1. Osip Mandelstam, *Tristia*, trans. Bruce McClelland (Station Hill, 1987), 41.

2. Ludwig Wittgenstein, *Philosophical Investigations*, trans. G. E. M. Anscombe (Blackwell, 1958), 223.

3. Aesop, *Aesop's Fables*, trans. L. Gibbs (Oxford University Press, 2002), 239. This translation is based on Babrius's version given in Edwin Perry, ed., *Babrius and Phaedrus* (Harvard University Press, 1965), 74–77.

4. The different version mentioned is Émile Chambry's and is provided in Edwin Perry, *Aesopica: A Series of Texts Relating to Aesop or Ascribed to Him or Closely Connected with the Literary Tradition That Bears His Name* (University of Illinois Press, 1952), 360, and has Prometheus as responsible for the making of the human.

5. Lucian, *Works*, vol. 6, trans. K. Kilburn (Harvard University Press, 1959), 297–99.

6. For a "short history" of transparency, its etymology, and its first semantic exploits, see Stefanos Geroulanos, "Transparency Dreaming Freedom (Maurice Blanchot's *The Most High*)," *MLN* 122, no. 5 (2007): 1050–78.

7. Liu Bolin's performative art is a case in point: the body of the artist perfectly camouflaged to match its surroundings disappears or, better, becomes invisibly transparent. H. G. Wells's *The Invisible Man* (1897) is a classic case of science fiction–generated transparency conceived as complete invisibility.

8. For an overview of cultural meanings associated with skin and its functions, see Stephen Connor, *The Book of Skin* (Cornell University Press, 2004).

9. Bernadette Wegenstein, *Getting Under the Skin: Body and Media Theory* (MIT Press, 2006), 6.

10. One broad cultural historical exploration of transparency that includes several figures of transparent humans within a wider range of transparency metaphors is Manfred Schneider, *Der Transparenztraum* (Matthes & Seitz, 2013). The book is

rich in literary and philosophical references and is organized around the idea of a dreamy and utopian quality of transparency discourse, from architecture to politics.

11. See Emmanuel Alloa and Dieter Thomä, eds., *Transparency, Society and Subjectivity: Critical Perspectives* (Palgrave Macmillan, 2018).

12. Walter Benjamin, "Surrealism: The Last Snapshot of the European Intelligentsia," trans. Edmond Jephcott, *NLR* 1, no. 108 (1978): 49. The passage continues: "It is also an intoxication, a moral exhibitionism, that we badly need. Discretion concerning one's own existence, once an aristocratic virtue, has become more and more an affair of petty-bourgeois parvenus" (49). On Benjamin's call for transparency, see also Emmanuel Alloa, "Transparency: A Magic Concept of Modernity," in Alloa and Thomä, *Transparency*, 21–56, 25. For a study of transparency and revolutionary policies in Russia between literature and politics see Julia Chadaga, *Optical Play: Glass, Vision, and Spectacle in Russian Culture* (Northwestern University Press, 2014). As for the glass architecture utopia, Chadaga examines the powerful counterexample provided by Eisenstein's *The Glasshouse* (1927–30), a radical project of critique of transparent social life as it had been achieved in America (162).

13. Anthony Vidler, *The Architectural Uncanny* (MIT Press, 1992), 217. On the tensions within architectural history between the utopia and the critique of glass, see Mechtild Widrich, " 'I'll Be Your Mirror': Transparency, Voyeurism and Glass Architecture," in *Invisibility Studies: Surveillance, Transparency and the Hidden in Contemporary Culture*, ed. Henriette Steiner and Kristin Veel (Peter Lang, 2015), 43–59.

14. Geroulanos, "Transparency Dreaming Freedom," 1074.

15. See Gianni Vattimo, *The Transparent Society* (Johns Hopkins University Press, 1992); and Jean Baudrillard, *The Transparency of Evil: Essays on Extreme Phenomena* (Verso, 1993). For Vattimo, the transparent society is a fiction perpetuated by communication systems (24–25); for Baudrillard, transparency is the stage where critical thought as reflection has ended: "Alienation is no more: the Other as gaze, the Other as mirror, the Other as opacity—all are gone. Henceforward it is the transparency of others that represents absolute danger. Without the Other as mirror, as reflecting surface, consciousness of self is threatened with irradiation in the void" (122).

16. See, e.g., Jane Bennett, *Vibrant Matter: A Political Ecology of Things* (Duke University Press, 2010); and Stacy Alaimo, *Bodily Natures: Science, Environment, and the Material Self* (Indiana University Press, 2010).

17. Rosi Braidotti, *Posthuman Knowledge* (Polity, 2019).

18. Alaimo, *Bodily Natures*, 4.

19. Arthur Lovejoy, *The Great Chain of Being: A Study of the History of an Idea* (Harvard University Press, 1950), 15.

20. I am using the notion of bodiliness echoing what Irina Hron has defined as "skinnedness," a critical term (modeled on Nietzsche's *Hautlichkeit*) to define the

resilient recurrence of skin in the artistic fantasies in which modifying skin aims nonetheless at transcending the body. See Irina Hron, "*Sit Venia Verbo*: A Case for Dermacriticism," *Orbis Litterarum* 79 (2024): 449–71.

21. On the social inscription of bodies, see Elizabeth Grosz, "Inscriptions and Body-Maps: Representations and the Corporeal," in *Feminine/Masculine and Representation*, ed. Terry Threadgold and Anne Cranny-Francis (Routledge, 2020), 62–74.

22. Donna Haraway, *When Species Meet* (University of Minnesota Press, 2008), 4.

23. On literature understood in terms of a heteronomous aesthetic, see Irene Albers, Marcus Hahn, and Frederic Ponten, eds., *Heteronomieästhetik der Moderne* (De Gruyter, 2022).

24. See Claudia Benthien, *Skin: On the Cultural Border Between Self and World*, trans. Thomas Dunlap (Columbia University Press, 2002); and Connor, *The Book of Skin*.

25. Questions of surveillance from a mostly architectural as well as media standpoint are prominent in Steiner and Veel's *Invisibility Studies*.

26. See Henriette Steiner and Kristin Veel, "Negotiating (In)Visibilities in Contemporary Culture: A Short Introduction," in Steiner and Veel, *Invisibility Studies*, xx.

Chapter 1

1. Alonso de Santa Cruz, *Sobre la Melancolía* (Ediciones Universidad de Navarra, 2005), 61 (my translation).

2. On the possible influence of this source, see Saturnino Rivera Manescau, "El modelo del Licenciado Vidriera," in *IV Centenario de Miguel de Cervantes* (Universidad de Valladolid, 1947), 3–11.

3. On melancholy as a European phenomenon, see Angus Gowland, "The Problem of Early Modern Melancholy," *Past & Present* 191, no. 1 (2006): 77–120, esp. 80.

4. Even the rhetorical presentation of these medical texts, which in their similar structures attest to the shared network of meanings within which the mental disorder was inscribed, can constitute an obstacle to the contemporary attempt to interpret the psychic disease as such. On the peculiar rhetoric of these texts, see Teresa Scott Soufas, *Melancholy and the Secular Mind in Spanish Golden Age Literature* (University of Missouri Press, 1990), 10. Soufas emphasizes the lack of originality of such texts, which boasted the repetition of a tradition and a hybrid form of composition, between "medical tract and artistic literature" (11). Gowland, in "The Problem of Early Modern Melancholy," speaks of an "instability" of such texts (83), pointing to the same difficulty in understanding them from the standpoint of contemporary expectations about medicine. In the text, I am using the forms "semeiology" and "semeiological" in accordance with the medical use.

5. Jean Starobinski, *L'Encre de la mélancolie*, ed. Fernando Vidal (Seuil, 2012), 15.

6. A notable exception to such anachronistic readings is the work by Gill Speak,

who advocates for integrating the short story into the philosophical and medical discourse of Cervantes's time, arguing against the tendency of many modern literary scholars to overlook how medical and philosophical writings shaped literature in the late sixteenth and early seventeenth centuries. See Gill Speak, "*El licenciado Vidriera* and the Glass Men of Early Modern Europe," *Modern Language Review* 85 (1990): 850–65, esp. 857. Speak has also written one more relevant article on the glass delusion, pointing to many of the historical medical sources that I also mention in this chapter: Gill Speak, "An Odd Kind of Melancholy: Reflections on the Glass Delusion in Europe (1440–1680)," *History of Psychiatry* 1 (1990): 191–206.

7. The summary of Cervantes's text and the quotations draw on the story's English translation in Miguel de Cervantes Saavedra, *The Complete Exemplary Novels*, ed. and trans. Barry Ife and Jonathan Thacker (Oxbow, 2013), 271–308. The original Spanish phrases or expressions are quoted from the Spanish version of the text in Miguel de Cervantes Saavedra, *Novelas ejemplares II*, ed. Francisco Rodríguez Marín (Espasa-Calpe, 1962), 9–83.

8. Cervantes, "The Glass Graduate," 283; Cervantes, "El licenciado Vidriera," 36.

9. While Speak insists on the melancholic symptomatology (Speak, "*El licenciado Vidriera*," 857), Alban Forcione dismisses the glass delusion in terms of the melancholic syndrome as something not pertinent to the actual narration of Vidriera—mainly because, unlike a melancholic, the glass graduate got his delusion from eating a charmed quince. See Alban K. Forcione, *Cervantes and the Humanist Vision: A Study of Four Exemplary Novels* (Princeton University Press, 1982), 274–75. Sybil Dümchen argues, from a different perspective, that the poisoned quince does not cause Vidriera's madness but intensifies a mental state he already experienced before his illness. See Sybil Dümchen, "The Function of Madness in *El licenciado Vidriera*," in *Cervantes's Exemplary Novels and the Adventure of Writing*, ed. Michael Nerlich and Nicholas Spadaccini (Prisma Institute, 1989), 99–123, esp. 104.

10. Hippocrates, *La Nature de l'homme*, ed. and trans. Jacques Jouanna (Akademie Verlag, 2002), 2–4.

11. In fact, later accounts of the melancholic humor compare it to an "ink," as in the expression "inchiostro" in Tommaso Campanella, *Il senso delle cose e della magia*, ed. Filiberto Walter Lupi (Rubbettino Editore, 2003), 153. On the metaphor of ink, see Starobinski, *L'Encre de la mélancolie*, 24.

12. Entirely dedicated to melancholy is Galen's *De atra bile*, but sections on melancholy are scattered across several other places of his vast opus. See Galen, "De atra bile," in Galen, *Claudii Galeni opera omnia*, vol. 5, ed. Karl Gottlob Kühn (Leipzig: Knobloch, 1823), 104–48.

13. On the persistence of humoral theory up to the close of the seventeenth century, and the following transformations within medicine, see Gowland, "The Problem of Early Modern Melancholy," 88. On the persistence of melancholy in Renaissance culture, and its dependency on ancient sources, see Roger Bartra, *El siglo de oro de la melancolía: Textos españoles sobre las enfermedades del alma* (Universidad Iberoamericana, Departamento de Historia, 1998), esp. 54–55.

14. Andrés Velásquez, *Libro de la melancolía* (M. Baroni, 2002). For a study of Velasquez's book, see Bartra, *El siglo de oro de la melancolía*.

15. Bartra emphasizes both the cultural provinciality of Velasquez's context and the merging of Galenic knowledge with Christian Scholasticism (see *El siglo de oro de la melancolía*, 13, 28).

16. Raymond Klibansky, Erwin Panofsky, and Fritz Saxl, *Saturn and Melancholy: Studies in the History of Natural Philosophy, Religion, and Art* (Nelson, 1964). For concrete examples of these alternative strands in the primary sources, see the theory of Ficino on melancholy as linked to astrology in Marsilio Ficino, *De vita libri tres / Drei Bücher über das Leben*, ed. Michaela Boenke (Wilhelm Fink, 2012), which was first published in 1489. For a theory of demonology as connected to melancholy, see Campanella, *Il senso delle cose e della magia*, published first in 1604. On the early modern history of demonology see Noel Brann, *The Debate over the Origin of Genius During the Italian Renaissance: The Theories of Supernatural Frenzy and Natural Melancholy in Accord and in Conflict on the Threshold of the Scientific Revolution* (Brill, 2002).

17. Rufus of Ephesus, *On Melancholy*, ed. Peter E. Pormann (Mohr Siebeck, 2008).

18. See Klibansky, Panofsky, and Saxl, *Saturn and Melancholy*, 87.

19. This shift implies diagnostic discrepancies with later uses of the same category. The phenomenon pertains to the history of psychiatry more at large, for example, to the history of schizophrenia as it has been critically reconstructed in Orna Ophir, *Schizophrenia: An Unfinished History* (Polity, 2022). In schizophrenia, delusions have always been central symptoms for a diagnosis (135), which indicates a symptomatic continuum with the diagnostic criteria that had been determining for melancholy.

20. Michel Foucault, *History of Madness*, ed. Jean Khalfa, trans. Jonathan Murphy and Jean Khalfa (Routledge, 2006), 265: "The morbid unity is not defined on the basis of observed signs or inferred causes, but half-way between the two, above some of them, it is perceived as a certain qualitative coherence, which has its own laws of transmission, development and transformation. It is the secret logic of this quality that orders the future of the notion of melancholy, not medical theory." Foucault insists on the symbolic unity that melancholy attains by way of juxtaposing symptoms, causes, and physiological processes: "A symbolic unity shaped by the sluggishness of the fluids, by the darkening of the animal spirits and the shadowy twilight that they spread over the images of things, by the viscosity of a blood that circulates through the vessels with difficulty, by the thickening of vapours that have blackened, acrid and deleterious, and by the slowed visceral functions, which become bogged down—more sensible than conceptual or theoretical, this unity gives melancholy its distinctive formula" (268).

21. See Drew Daniel, *Melancholy Assemblage: Affect and Epistemology in the English Renaissance* (Fordham University Press, 2013). Daniel ascribes the fluidity of melancholy's meanings to a "discursive surplus" that determines melancholy's "crowded scene" (5). In similar terms, Roger Bartra has spoken of early modern mel-

ancholy's "thick sentimental texture" (Bartra, *El siglo de oro de la melancolía*, 30). Proof to the pervasive appeal of the notion is the extraordinary historical persistence of the word *melancholy* through continuous reconfigurations of meanings, which Jean Starobinski has partly explained with a fundamental taste for verbal continuities (Starobinski, *L'Encre de la mélancolie*, 16).

22. Galen, "De locis affectis Liber III," in Galen, *Claudii Galeni opera omnia*, vol. 8, 136–215, 190.

23. Galen, "Definitiones medicae," in Galen, *Claudii Galeni opera omnia*, vol. 19, 346–462, 416.

24. Rufus of Ephesus, *On Melancholy*.

25. See Soufas, *Melancholy and the Secular Mind*, 21–23; and Gowland, "The Problem of Early Modern Melancholy," 114 and note 141.

26. Aristotle, "Book 30," in *Problems*, vol. 2, Books 22–38, trans. W. S. Hett (Harvard University Press, 1957), 154–81. The text was long attributed to Aristotle but, in fact, was a collection of spurious texts by the peripatetic school.

27. Klibansky, Panofsky, and Saxl, *Saturn and Melancholy*, 68.

28. Enea Silvio Piccolomini, *I commentarii*, 2 vols., ed. Luigi Totaro (Adelphi, 2008), 1056. The work was written between 1462 and 1463 but later censored for print. The anecdote goes: "He sometimes believed himself to be made of glass, and he would not allow himself to be touched. He inserted iron rods into his clothing and armed himself in many ways so that, if he fell, he would not be broken. He had, however, lucid intervals, and when he returned to his senses, he was ashamed of his former life; for that reason, he loved solitude" (my translation).

29. Levinus Lemnius, *Levini Lemnii de habitu et constitutione corporis* (Frankfurt: N. Hofmann for I. Rhodius, 1604), 180 (my translation). An English translation was already available in 1576 under the title *Touchstone of Complexions*.

30. André Du Laurens, *A Discourse of the Preservation of the Sight: Of Melancholike Diseases; Of Rheumes; And of Old Age*, trans. Richard Surphlet (London: Felix Kingston for Ralph Iacson, 1599), 102. The text is an early English translation of the original French.

31. "A glassmaker in Paris, living in the suburb of Saint-Germain, believed that he had buttocks made of glass, and even when standing upright he always protected them with a cushion swollen with many feathers. By order of the doctors he was taken away from the glass workshop and beaten with rods. And when, under the blows, he groaned, he was told that he was feeling fleshly buttocks, not glass ones. Thus he came to his senses. Another man, claiming that he had feet made of glass, walked with great caution." Ludovicus Casanova [Louis de Caseneuve], *Hieroglyphicorum et medicorum emblematum* (Lugduni [Lyon]: Sumptibus Pauli Frellon, 1626), 58 (my translation). On the emblematic format of Caseneuve's medical work, see Magdalena Kozluk, "Une *imaginotheca* curieuse: Les *emblemata medica* de Louis de Caseneuve," *History of Science and Medicine* 50, no. 3 (July 2016): 277–88.

32. On the European circulation, see Gowland, "The Problem of Early Modern Melancholy."

33. The delusion is, for example, not mentioned in the highly popular 1586 English handbook, probably also known to Shakespeare, by Timothie Bright, *A Treatise of Melancholie* (Facsimile Text Society by Columbia University Press, 1940).

34. René Descartes, *Meditations on First Philosophy: With Selections from the Objections and Replies*, ed. and trans. John Cottingham (Cambridge University Press, 1986), 13. "Again, how could it be denied that these hands or this whole body are mine? Unless perhaps I were to liken myself to madmen, whose brains are so damaged by the persistent vapours of melancholia that they firmly maintain they are kings when they are paupers, or say they are dressed in purple when they are naked, or that their heads are made of earthenware, or that they are pumpkins, or made of glass. But such people are insane, and I would be thought equally mad if I took anything from them as a model for myself." This passage elicited a critical exchange between Foucault and Derrida on how to read the presence of madness in Descartes's philosophy. See Michel Foucault, "My Body, This Paper, This Fire," trans. Geoff Bennington, *Oxford Literary Review* 4, no. 1 (Autumn 1979): 9–28. For Foucault, Descartes operates a fundamental exclusion of madness from rational subjectivity: "the meditating subject had to exclude madness by qualifying himself as not mad" (24)—an exclusion that Derrida's reading would instead obscure.

35. Burton's work can be located within a tradition of moral and spiritual approaches to melancholy. On such elements in Burton, see Gowland, "The Problem of Early Modern Melancholy," 107; and Mary Ann Lund, *Melancholy, Medicine and Religion in Early Modern England* (Cambridge University Press, 2010), 171.

36. Robert Burton, *The Anatomy of Melancholy*, ed. Thomas C. Faulkner, Nicolas K. Kiessling, and Rhonda L. Blair (Oxford University Press, 1989), part 1, sec. 3, memb. 1, subsec. 2, 386.

37. Burton, *The Anatomy of Melancholy*, part 1, sec. 3, memb. 1, subsec. 3, 403.

38. Burton, *The Anatomy of Melancholy*, part 1, sec. 3, memb. 2, subsec. 1, 410.

39. I am not referring here to the diagnostic category of body dysmorphic disorder as listed in the DSM-5, which would be yet another anachronism forced on the early modern syndrome.

40. Tomaso Garzoni, *L'ospidale dei pazzi incurabili*, in *Opere*, ed. Paolo Cherchi (Longo Editore, 1993), 267–68.

41. Du Laurens, *A Discourse*, 101.

42. Du Laurens, *A Discourse*, 102.

43. Du Laurens, *A Discourse*, 103.

44. Du Laurens, *A Discourse*, 603. As is to be expected from the early modern textual circulation and formatting of texts, in Caseneuve's *Emblemata* quoted above, from 1626, these anecdotical cases return in quite the same fashion.

45. At play here is what Bernadette Wegenstein calls the "struggle between holism and fragmentation" as the main axis along which the history of the body unfolds. See Bernadette Wegenstein, *Getting Under the Skin: Body and Media Theory* (MIT Press, 2006), 35.

46. I am using the notion of glitch with the political potential ascribed to it by Legacy Russell, *Glitch Feminism: A Manifesto* (Verso, 2020).

47. Cervantes, "El licenciado Vidriera," 36.

48. Cervantes, "El licenciado Vidriera," 36.

49. Cervantes, "The Glass Graduate," 283; "El licenciado Vidriera," 37.

50. Cervantes, "The Glass Graduate," 291; "El licenciado Vidriera," 48–49.

51. Cervantes, "El licenciado Vidriera," 53, 56.

52. Cervantes, "The Glass Graduate," 293; "El licenciado Vidriera," 58.

53. Cervantes, "The Glass Graduate," 296; "El licenciado vidriera," 61. The pun is based on the custom of sitting on the shoulders of the condemned to finish them off.

54. Cervantes, "The Glass Graduate," 297; "El licenciado vidriera," 62. The pun here plays with the double meaning of *cuento* as story and money account.

55. Cervantes, "The Glass Graduate," 301. "Aunque de vidrio, no soy tan frágil que me deje ir con la corriente del vulgo, las más veces engañado" (Cervantes, "El licenciado Vidriera," 71).

56. Cervantes, "The Glass Graduate," 283; "El licenciado Vidriera," 37.

57. "In those who thought they had no head, or that they saw black men, the 'imaginatio' was disturbed, while understanding and memory remained intact" (Klibansky, Panofsky, and Saxl, *Saturn and Melancholy*, 93).

58. On the three levels of the soul and their complex distribution, see Miranda Anderson, *The Renaissance Extended Mind* (Palgrave Macmillan, 2015), 87. On Aquinas's understanding and structuring of the sensitive soul, see Anthony J. Lisska, *Aquinas's Theory of Perception: An Analytic Reconstruction* (Oxford University Press, 2016).

59. In fact, integral to the imagination or *vis imaginativa* is the sensory apparatus or *sensus communis*. For a detailed account of this mapping in Aristotle, Avicenna, and Aquinas, see Lisska, *Aquinas's Theory*, 215–18.

60. See Juan Huarte, *Examen de ingenios para las ciencias*, ed. Guillermo Serés (Cátedra, 1989), 206: "memoria, imaginativa y entendimiento." See also Du Laurens, *A Discourse*, 101; and Garzoni, *L'ospidale dei pazzi incurabili*, 267.

61. Soufas, *Melancholy and the Secular Mind*, 17.

62. As for the notion of *ingenio*, Cervantes was following a standard notion adopted, for example, by philosopher Juan Luis Vives in *De anima et vita* (1538), who defined *ingenium* as "universa nostrae mentis vis"—that is, "the whole force of our mind." See Juan L. Vives, *De anima et vita*, ed. and trans. Mario Sancipriano (Bottega d'Erasmo, 1963), 167. Another comparable notion of *ingenio* is implemented by the physician Juan Huarte de San Juan in his *Examen de ingenios para las ciencias* (1578). This Spanish doctor had articulated a complex natural philosophy of human *ingenio*, which he conceived as the totality of mental faculties, on the one hand, and the individually unique intellectual talent, on the other. The direct influence of Huarte on Cervantes is debated: see Winfried Schleiner, *Melancholy, Genius, and Utopia in the Renaissance* (Harrassowitz, 1991), 166–67, but the shared

medical knowledge is evident (see Malcolm K. Read, *Juan Huarte de San Juan* (Twayne, 1981), 110–14. On the influence of Huarte on the *Don Quixote*, see, in particular, Otis Green, "El ingenioso Hidalgo," *Hispanic Review* 25, no. 3 (1957): 175–93. On Cervantes and Huarte possibly meeting in 1591 in Baeza, see Carroll B. Johnson, *Madness and Lust: A Psychoanalytical Approach to Don Quixote* (University of California Press, 1983), 18.

63. Cervantes, "The Glass Graduate," 283; Cervantes, "El licenciado Vidriera," 37.

64. As to the peculiar continuity of the glass delusion into the present, see Andy Lameijn, "Le corps de verre: Histoire d'un délire prototypique," in *Frénésie, histoire psychiatrie psychanalyse II* 10 (1992): 205–30. As to the readings of the story, many a reader has become entangled in an anachronistic symbology. As cases in point, see Ruth S. El Saffar, *Novel to Romance: A Study of Cervantes's Novelas Ejemplares* (Johns Hopkins University Press, 1974); Forcione, *Cervantes and the Humanist Vision*; Dümchen, "The Function of Madness"; Stanislav Zimic, "El licenciado Vidriera: La tragedia del intelectual integro," *La Torre* 6, no. 22 (1992): 237–70; and Stephen Rupp, "Soldiers and Satire in *El licenciado Vidriera*," in *A Companion to Cervantes's Novelas Ejemplares*, ed. Stephen Boyd (Tamesis, 2005), 134–47. Partly leaning toward anachronism in the suggestion of a "schizophrenic structure" is also Cesare Segre, "La struttura schizofrenica del 'Licenciado Vidriera' di Cervantes," in *Fuori del mondo: I modelli nella follia e nelle immagini dell'aldilà* (Einaudi, 1990). Alban Forcione's reading, which is otherwise very compelling, is paradigmatic of the anachronistic tendency of reading the story: "the glass of his delusion is in fact symbolic of the keen insight with which he is gifted and the freedom of a man totally dedicated to the spirit" (241). He also argues that "the scholar's fanatical devotion to the mind and the spirit is grotesquely caricatured in his removal from his body and in the terribly high price which, in his state of constant anxiety, he must pay for his knowledge" (242). The glass delusion is hence read by Forcione mostly as a metaphor, be it of knowledge through the semantics of vision or of objectification and dehumanization through a certain understanding of the inorganic matter of glass. Several readers have repeated, with some variations, this reading of Vidriera's delusion. For Rupp, "the particular affliction that the text describes has explicit associations with the transparency and keenness of satirical discourse" (134). For Zimic, Vidriera's "delusion suggests both the clarity and the precariousness of truth in society" (137). El Saffar goes further in allowing a metaphor forced onto the text to generate an interpretive cascade: "Tomás' trip through Italy becomes a travelogue in which his personality becomes totally insignificant. He becomes pure transparency, simply conveying reactions and impressions" (54). Dümchen, in similar fashion, claims that Tomás's glass delusion is a "fantasy" of "protection," which "saves him from having to take into account different aspects of life unrelated to rationalism" (119).

65. See the interesting remarks by Forcione in *Cervantes and the Humanist Vision*, 315–16 and n178. Forcione insists on the disembodiment of Vidriera and sees it as a punitive form of *contrapasso* for the sins of a "misdirected" scholar (305).

66. Cervantes, "The Glass Graduate," 283; "El licenciado Vidriera," 36.

67. Cervantes, "The Glass Graduate," 285; "El licenciado Vidriera," 38.

68. Cervantes, "The Glass Graduate," 285; "El licenciado Vidriera," 39.

69. Cervantes, "The Glass Graduate," 303; "El licenciado Vidriera," 75.

70. "Preguntóle uno que cómo sentía aquella avispa, si era su cuerpo de vidrio. Y respondió que aquella avispa debía de ser murmuradora, y que las lenguas y picos de los murmuradores eran bastantes a desmoronar cuerpos de bronce, no que de vidrio" (Cervantes, "El licenciado Vidriera," 75). The polemics against slanderers is a topos of early modern Christian ethics. Erasmus, in the first chapter of his Christian "handbook" *Enchiridion*, uses a similar image to that of the sting: "Thy body is not alive if it feel not the pricking of a pin. And is thy soul alive which lacketh the feeling of so great a wound? Thou hearest some man use lewd and presumptuous communication, words of backbiting, unchaste and filthy, raging furiously against his neighbour: think not the soul of that man to be alive." Desiderius Erasmus, *A Book Called in Latin "Enchiridion Militis Christiani" and in English "The Manual of the Christian Knight"* (Methuen, 1905), 51. If we take Erasmus's passage as a diagnostic guide for detecting the health condition of soul and body, we should infer that Vidriera's body is, in fact, healthy.

71. Forcione, here too, reads this scene as a striking caricatural scene of *contrapasso*: "the fool's mask drops, and we watch him as he complains about the very pain that he continually inflicts on others" (Forcione, *Cervantes and the Humanist Vision*, 281). George Shipley has objected to this reading, based on a fundamentally negative assessment of the licentiate's life as an individualist, self-involved scholar, by affirming the positive nature of Tomás Rodaja's friendships and interactions, lifting the weight of any *contrapasso* from the "inhuman" scholar depicted by Forcione and others. See George A. Shipley, "Garbage In, Garbage Out: 'The Best of Vidriera,'" *Cervantes: Bulletin of the Cervantes Society of America* 21, no. 1 (2001): 5–41.

72. Maurice Merleau-Ponty, *Phenomenology of Perception*, trans. Donald A. Landes (Taylor & Francis, 2013), 213. Merleau-Ponty's phenomenology offers a viable set of philosophical tools to take Vidriera's embodiment (and that of his fellow glass humans) seriously. For the glass body, it might be meaningful to adopt the notion of the body as "a system of motor powers or perceptual powers": "not an object for an 'I think'" but "a totality of lived significations that moves toward its equilibrium" (155).

73. For a global history of glass see Alan Macfarlane and Gerry Martin, *Glass: A World History* (University of Chicago Press, 2002).

74. Galen, "De locis affectis Liber III," 190: "Siquidem alius testaceum se factum putavit": "Likewise another believed himself made of clay" (my translation).

75. Harald Weinrich, *Das ingenium Don Quijotes* (Aschendorf, 1956), 51.

76. Ecclesiastes 12:6–8: "Before the silver cord is snapped, and the golden bowl is broken, and the pitcher is broken at the fountain, and the wheel broken at the cistern, and the dust returns to the earth as it was, and the breath returns to God who gave it. Vanity of vanities, says the Teacher; all is vanity." *New Oxford Annotated*

Bible: New Revised Standard Version: An Ecumenical Study Bible, 5th ed., ed. Michael D. Coogan (Oxford University Press, 2018); hereafter NOAB (NRSV).

77. Job 10:8–9, NOAB (NRSV); Job 33:6, NOAB (NRSV); Isaiah, 64:8, NOAB (NRSV).

78. 2 Corinthians 4:7–10, NOAB (NRSV).

79. Cervantes, "The Glass Graduate," 285; "El licenciado Vidriera," 37.

80. On the urinal motif, see Speak, "*El licenciado Vidriera*," 858. In the story Vidriera is even once addressed as "señor Redoma" (Mr. Flask).

81. "Tutti siamo di vetro e, sottoposti ad innumerabili pericoli, per piccola sospinta siamo rotti e torniamo in nulla." Letter to Francesco Nelli, priore at Sant'Apostolo in Certaldo, later serving for the Gran Siniscalco in Naples Niccolo' Acciajuoli; Giovanni Boccaccio, *Opere minori in volgare*, ed. Mario Marti, vol. 4 (Rizzoli, 1972), 761 (my translation). On the Latin use of *vitreus* as referring to the fragility of fame or fortune, see Alfred Garvin Engstrom, "The Man Who Thought Himself Made of Glass, and Certain Related Images," *Studies in Philology* 67, no. 3 (1970): 390–405, 403.

82. Vannoccio Biringuccio, *De la pirotechnia: Libri X* (Venezia: Venturino Roffinello, 1540), 44 (my translation).

83. The original Latin passage is in Casanova, *Emblemata*, 8. The excerpt presented here is my translation.

84. In one variant of the glass's brittleness allegory of life, it is women who are seen as similar to glass objects, mostly, however, due to their alleged vanity and their being bound to glass mirrors, as argued in Anderson, *The Renaissance Extended Mind*, 189–96.

85. On the "exceptionality" of developing clear glass from a global historical perspective, see Macfarlane and Martin, *Glass*, 119.

86. The suggestion of the general melancholic photophobia is contained in Speak, "An Odd Kind of Melancholy," 197–201. Speak follows an interesting interpretative line identifying the positive value of transparency in mystical writings, from Santa Teresa to San Juan de la Cruz to Giordano Bruno. My reading of the transparency of glass melancholics relies on different sources and leads in a quite different direction.

87. The whole passage reads: "È assai ridicoloso ancora quello [esempio] di colui che, parendoli esser divenuto un vetro, andò a Murano per gettarsi dentro a una fornace e farsi fare in foggia d'una inghistara." Garzoni, *Il teatro de' vari e diversi cervelli mondani*, in *Opere*, 216 (my translation).

88. The Italian *inghistara* or *angastara*, from the Greek ἄγγος (vessel) and γαστήρ (belly), indicates a vessel with a tall neck. For historical details regarding the *inghistara*, see Rosa Barovier Mentasti and Cristina Tonini, "Tools to Study Glass: Inventories, Paintings and Graphic Works of the 16th Century," in *Study Days on Venetian Glass: Approximately 1600's*, ed. Rosa Barovier Mentasti and Cristina Tonini (Istituto Veneto di Scienze, Lettere ed Arti, 2014), 3–34, 7.

89. Garzoni is one more historical source for the preeminence of Murano in the sixteenth century. In yet another encyclopedic compendium dedicated to professions and crafts, he delivers a section on glassmaking focusing on the exceptionality of Murano. The passage is not included in the edition of the *Opere* quoted so far. Tommaso Garzoni: *La piazza universale di tutte le professioni del mondo* (Venezia: Somasco, 1593): "Ma oggidì Murano luogo eminissimo e deliziosissimo presso a Venezia supera tutti i luoghi del mondo di vetri, e di christalli" (540).

90. A comparison between some of Maggi's designs and actual objects, devoted to the bottle called *Kuttrolf*, is in Mauro Stocco, "Kuttrolf seicenteschi dal Museo del Vetro di Murano," in *Aldèbaran* VII: *Storia dell'arte*, ed. Sergio Marinelli (Scripta Edizioni, 2025), 57–78.

91. For a history of Murano and its association with the production of glass, see Marco Toso Borella, "Il vetro," in *Atlante storico di Venezia*, ed. Giovanni Distefano (Supernova Edizioni, 2007), 1059–64.

92. Marco Verità, "Secrets and Innovations of Venetian Glass Between the 15th and the 17th Centuries: Raw Materials, Glass Melting and Artefacts," in Mentasti and Tonini, *Study Days on Venetian Glass*, 53–63. On the ingredients of Muranese glass, see also Toso Borella, "Il vetro," 1061.

93. I am grateful to Michele Zampedri for introducing me to some Muranese technical vocabulary and to some glassmakers in the furnaces of Murano. I am especially grateful to Stefano Dalla Valentina for showing me the process of making a glass woman. Marco Toso Borella has provided me with more firsthand knowledge on glassmaking in Murano between past and present. For the effects of the free hand technique, see also Astone Gasparetto, "Dalla realtà archeologica a quella contemporanea," in *Mille anni di arte del vetro a Venezia*, ed. Rosa Barovier Mentasti, Attilia Dorigato, Astone Gasparetto, and Tullio Toninato (Albrizzi, 1982), 15–38, 25.

94. Jutta-Annette Page, *Beyond Venice: Glass in Venetian Style, 1500–1750* (Corning Museum of Glass, 2004), 6; Verità, "Secrets and Innovations," esp. 55 on the chemical composition of Muranese glass.

95. Antonio Neri, *L'arte vetraria*, ed. Ferdinando Abbri (Giunti, 2001), 38, 42. For a meticulous reconstruction of Neri's life, see Paul Engle. *Conciatore: The Life and Times of 17th Century Glassmaker Antonio Neri* (Heiden & Engle, 2014). Recipes for glassmaking were circulating also before, and some manuscripts have remained; they were, however, kept secret and transmitted within the single furnaces led by different masters (*maestri*). On the first documents left about the making of cristallo, see Luigi Zecchin, *Vetro e vetrai di Murano*, vol. 1 (Arsenale Editrice, 1987), 229–33.

96. Verità, "Secrets and Innovations," 57–58. On Barovier and the concrete innovations in the making of *cristallo*, which included several washings of the frits (*fritta*) in water, see Gasparetto, "Dalla realtà archeologica a quella contemporanea," 22.

97. Giovanni Mariacher, *I vetri di Murano* (Carlo Bestetti Edizioni d'Arte,

1967), 31. Even Galileo built his telescope using a glass lens produced in Murano; see Verità, "Secrets and Innovations," 58.

98. Mariacher, *I vetri di Murano*, 41.

99. The body technique of the glassblower is an acquired practical knowledge that requires both technique and strength, as it is possible to observe today still when visiting a furnace.

100. John Locke, *An Essay Concerning Human Understanding*, ed. Peter H. Nidditch (Oxford University Press, 1979), 325.

101. See Annette Cremer, "Glas. Faszinierendes Material der Eliten in der Frühen Neuzeit," in *Glas in der Frühen Neuzeit Herstellung: Verwendung, Bedeutung, Analyse, Bewahrung* (Heidelberg University Publishing, 2022), 16–81, 27.

102. See Zecchin, *Vetro e vetrai*, 233.

103. Ignasi Domenéch, "Spanish Façon de Venise Glass," in Page, *Beyond Venice*, 84–141, 85. Domenéch specifies that Muranese cristallo glasses became essential tableware in Catalonia, and among aristocratic and cultured circles, they also became collectible items and markers of social prestige (87). The popularity of Venetian glass, however, started to decline toward the end of the seventeenth century, when its dominance was surpassed by French, Bohemian, and English glass (see Marie-Laure de Rochebrune, "Venetian and Façon de Venise Glass in France in the 16th and 17th Century," in Page, *Beyond Venice*, 142–91, 146).

104. Cremer, "Glas. Faszinierendes Material," 16–17.

105. Cervantes, "The Glass Graduate," 283; "El licenciado Vidriera," 37.

106. Platonic philosophy of the soul as formulated in the *Phaedo*, for example, defined the body as a garment for the soul, a temporary dwelling, and an instrument. Plato, *Phaedo*, in *Euthyphro, Apology, Crito, Phaedo, Phaedrus*, trans. Harold North Fowler, intro. by W. R. M. Lamb (Harvard University Press, 1914), 303–5. The instrumental function of the body in relation to the soul was also argued by Aristotle but within a different set of questions and with different purposes. Notoriously, the main idea of Aristotle's *De anima* is that the soul is the formal, organizational principle for the life of the body, suggesting a much more dynamic relation between soul and body than that of Platonism.

107. "The body is constantly breaking in upon our studies and disturbing us with noise and confusion, so that it prevents our beholding the truth, and in fact we perceive that, if we are ever to know anything absolutely, we must be free from the body and must behold the actual realities with the eye of the soul alone." Plato, *Phaedo*, 231.

108. Plato, *Phaedo*, 233.

109. Plato, *Phaedo*, 290–91. The Greek verb indicates literally the captivity of the soul under the body: καταδεῖται ψυχὴ ὑπὸ σώματος.

110. Plato, *Phaedrus*, in *Euthyphro, Apology, Crito, Phaedo, Phaedrus*, 484–85.

111. Augustine, *Confessions and Enchiridion*, ed. Albert Cook Outler (Westminster Press, 1955). In the *Confessions*, the image of dwelling and of the house occurs often, although not strictly in relation to the soul/body nexus; for example: "The

house of my soul is too narrow for thee to come in to me; let it be enlarged by thee. It is in ruins; do thou restore it. There is much about it which must offend thy eyes; I confess and know it. But who will cleanse it? Or, to whom shall I cry but to thee?" (34; see also 151, 169). Meanwhile, *City of God* is teeming with images of dwelling in the "house of God." Augustine, *Concerning the City of God Against the Pagans*, trans. Henry Bettenson, intro. by G. R. Evans (Penguin, 2003). *City of God* abounds in remarks on the dualism of soul and body but also on the productive interdependency of body and mind: "If you have enough light in your minds to prefer the soul to the body, choose which you should worship!" And: "the body, which is the servant of the mind, . . . the mind itself, the actual ruler of the flesh" (44, 53).

112. The trope is crucial to Christian Neoplatonism in general. See Erasmus: "A man is then a certain monstrous beast, compact together of two parts. . . . If thy body had not been added to thee, thou hadst been a celestial or godly thing" (*A Book Called in Latin "Enchiridion,"* 81).

113. Juan L. Vives, *De anima et vita*, ed. and trans. Mario Sancipriano (Bottega d'Erasmo, 1963), 199. On the metaphor of *habitare*, see also the brief remarks by Sancipriano (203n1).

114. Vives, *De anima et vita*, 217.

115. Vives, *De anima et vita*, 177, 333, 339, 417.

116. Vives, *De anima et vita*, 257.

117. Vives, *De anima et vita*, 286–87 (my translation, here and after).

118. Vives, *De anima et vita*, 441.

119. Vives, *De anima et vita*, 275.

120. Vives, *De anima et vita*, 297.

121. For an iconological reading of Dürer's engraving, see Klibansky, Panofsky, and Saxl, *Saturn and Melancholy*, part 4.

122. Vives, *De anima*, 294.

123. Ficino, *De vita libri tres*, 58.

124. Rosi Braidotti, *Posthuman Knowledge* (Polity, 2019).

125. Kevin LaGrandeur, "Early Modern," in *The Cambridge Companion to Literature and the Posthuman*, ed. Bruce Clarke and Manuela Rossini (Cambridge University Press, 2017), 16–28, 16.

126. Klibansky, Panofsky, and Saxl *Saturn and Melancholy*, 68; for an exegesis of Aristotle, 17–41, especially for a discussion on the exceptionality of the melancholics (31). On Ficino and the representation of melancholy, see Elena Fabietti, "Die Farbenlehre der Melancholie Marsilio Ficinos Beitrag zur frühneuzeitlichen Darstellung der Melancholie," *Schriftstücke* 3 (2020): 37–59.

127. Ficino, *De vita libri tres*, 55 (my translation).

128. Plato, *Phaedo*, 225.

129. Ficino is repeating and reinventing a classical tradition of thought that associated strenuous studies with the manifestation of melancholy. See, for example, the case history in Rufus, *On Melancholy*, 69, which complements the pseudo-

Aristotelian explanation by suggesting that excellent natures are predisposed to melancholy because they "move quickly and think a lot" (47).

130. Ficino, *De vita libri tres*, 57 (my translation). Ficino repeated the already pre-Galenic (i.e., apparent already in Rufus) division seen above of black bile into a natural humor and an unnatural one, the latter being a product of combustion of humors, often referred to as *melancholia adusta*. He therefore distinguished between bad and good melancholy using this same distinction, and assigning "goodness" to natural black bile only if it exist in the perfect proportion in the blood—a mixture that, quite surprisingly, would be purple and gold in color (62).

131. Huarte, *Examen*, 281–82 (my translation).

132. Katherine Hayles, *How We Became Posthuman: Virtual Bodies in Cybernetics, Literature, and Informatics* (University of Chicago Press, 1999), 4.

133. Cervantes, "The Glass Graduate," 303; "El licenciado Vidriera," 75.

134. Psalms 105:15, NOAB (NRSV); 1 Chronicles 16:22, NOAB (NRSV).

135. Here I am loosely referring to Agamben's definition of *homo sacer* as absolutely killable but unsacrificeable. The analogies I intend to draw, however, end here, as the problem of political power exceeds the scope of my inquiry. See Giorgio Agamben, *Homo sacer: Il potere sovrano e la nuda vita* (Einaudi, 2005), 81–82.

Chapter 2

1. For an encompassing assessment of the "visual primacy" of the Enlightenment, though "by no means without its complications," see Martin Jay, *Downcast Eyes: The Denigration of Vision in Twentieth-Century French Thought* (University of California Press, 1993), 103. Jay also argues that "the mainstream scopic regime of the Enlightenment" may have been only "apparent" (105). On the ambiguous "visual primacy" of Enlightenment culture, see also Barbara Maria Stafford, "Voyeur or Observer? Enlightenment Thoughts on the Dilemmas of Display," *Configurations* 1, no. 1 (1993): 95–128. On the "project of visibility" pursued by Enlightenment science, but first of all optics, see Theresa Levitt, *The Shadow of Enlightenment. Optical and Political Transparency in France, 1749–1848* (Oxford University Press, 2009), 2.

2. The description of an "anthropologische Konjunktion" is in Albrecht Koschorke, *Körperströme und Schriftverkehr: Mediologie des 18. Jahrhunderts* (Wilhelm Fink, 2003), 9.

3. The title's translation into English is "Physiognomical Fragments for the Promotion of the Love and Understanding of Mankind." I speak here of revamping because physiognomy as a science has a much longer history, which already had a flourishing phase in print form from the end of the fifteenth century. For a thorough reconstruction of early modern physiognomy, see Martin Porter, *Windows of the Soul: The Art of Physiognomy in European Culture, 1470–1780* (Clarendon Press, 2005).

4. As to Rousseau's and Sterne's acquaintance with each other, the two never met but were certainly aware of each other thanks to mutual friends. See Laurence

Sterne, *The Life and Opinions of Tristram Shandy, Gentleman*, in *The Florida Edition of the Works of Laurence Sterne*, ed. Melvyn New and Joan New, vol. 2 (University Presses of Florida, 1978), 538; and Wilbur L. Cross, *The Life and Times of Laurence Sterne* (Macmillan, 1909), 343. For the reception of Sterne in France at the end of the nineteenth century in relation to Rousseau, see Peter Jan de Voogd and John Neubauer, *The Reception of Laurence Sterne in Europe* (Continuum, 2004), 63.

5. The rhetorical register of the work hinges on the notion of *Mitleid* (compassion), which had a prominent role in the literature of *Empfindsamkeit* (sensibility) of the late Enlightenment; see Stephanie Bölts, *Krankheiten und Textgattungen: Gattungsspezifisches Wissen in Literatur und Medizin um 1800* (De Gruyter, 2016), 174–206, esp. 175–76.

6. Christian Heinrich Spieß, *Biographien der Wahnsinnigen*, ed. Walter Promies (Luchterhand, 1976), 48. In the text, when I provide an English translation of the German text, it is always mine.

7. Spieß, *Biographien der Wahnsinnigen*, 50.

8. "Vor dir verberge ich mein Herz nicht." Spieß, *Biographien der Wahnsinnigen*, 52.

9. "Da wurde meine Brust zur Laterne!" And later on: "Ist hier und hier nicht alles von Glas, so durchsichtig, als ob's Kristall wäre? Siehst du mein Herz und alle meine Gedanken darinnen?" Spieß, *Biographien der Wahnsinnigen*, 52–53.

10. "Wie leicht stockt das dann immer laufende Rad unsrer Einbildungskraft, welches gleich einer Laterna magika die Bilder der Vergangenheit und Zukunft vor unserer Seele vorüber dreht." Spieß, *Biographien der Wahnsinnigen*, 56.

11. "Öffnet meinen Sarg nicht, damit die Leute nicht in meinem Herzen lesen können!" Spieß, *Biographien der Wahnsinnigen*, 60.

12. "Schäme dich, ungeratenes Kind . . . , willst du mir denn nichts als Schande machen? Du hast eine Brust von Glase. . . . Verstecke dich!" Spieß, *Biographien der Wahnsinnigen*, 114.

13. Shame is theorized in phenomenological terms in Levinas's *De l'évasion* (1935): Emmanuel Levinas, *On Escape*, trans. Bettina Bergo, introduced and annotated by Jacques Rolland (Stanford University Press, 2003) and in Sartre's *L'Être et le néant* (1943): Jean-Paul Sartre, *Being and Nothingness*, trans. by Hazel Barnes (Routledge, 2003). A classic psychoanalytic analysis is provided in Léon Wurmser, *The Mask of Shame* (Johns Hopkins University Press, 1981). On the connection between shame and gaze, see Claudia Benthien, *Tribunal der Blicke. Kulturtheorien von Scham und Schuld und die Tragödie um 1800* (Böhlkau Verlag, 2011), esp. 89. On the psychoanalytical affect theory of shame, see Achim Geisenhanslücke, "Scham: Theorie und Geschichte einer starken Empfindung," in *Freiburger literaturpsychologische Gespräche* 32, ed. Joachim Küchenhoff, Joachim Pfeiffer, and Carl Pietzcker (2013): 21–40, 28. For a moral exploration of shame, see Martha C. Nussbaum, *Hiding from Humanity: Disgust, Shame, and the Law* (Princeton University Press, 2004). For a historical account of the rich debate around notions of guilt and shame

and their connection in light of psychoanalysis, see Ruth Leys, *From Guilt to Shame: Auschwitz and After* (Princeton University Press, 2007).

14. On the social semiotics of shame, see Alexandra Pontzen and Heinz-Peter Preußer, "Ritualisierte Verarbeitungsformen von Fehlverhalten. Eine Einleitung zu *Schuld und Scham*," in *Schuld und Scham* (Winter, 2008), 7–24, 15.

15. On the genre of Spieß's book, see Bölts, *Krankheiten und Textgattungen*.

16. Levinas, *On Escape*, 65: "Shame is, in the last analysis, an existence that seeks excuses. What shame discovers [*découvre*] is the being who uncovers himself [*se découvre*]." Levinas was tracing the possibility of the ontological subject of Western philosophy to escape its own cage. Giorgio Agamben has glossed the passage: "In shame, the subject . . . has no other content than its own desubjectification; it becomes witness to its own disorder, its own oblivion as a subject." Giorgio Agamben, *Remnants of Auschwitz: The Witness and the Archive* (Zone, 1999), 106.

17. Hence the challenge of a theory of shame, as soon as cultural analysis is opened up: see Michaela Bauks and Martin F. Meyer, "Einleitung," in *Zur Kulturgeschichte der Scham*, ed. Michaela Bauks and Martin F. Meyer (Felix Meiner Verlag, 2011), 7–15, esp. 15.

18. Alexander Košenina, "Gläserne Brust, lesbares Herz: Ein psychopathographischer Topos im Zeichen physiognomischer Tyrannei bei C. H. Spieß und anderen," *German Life and Letters* 52, no. 2 (1999): 151–65, in particular on the text as a reaction to physiognomy (155, 157, 165). Eduard in Goethe's *Wahlverwandtschaften* (1809) and Nathanael in E.T.A. Hoffmann's *Der Sandmann* (1817) are two prominent examples provided by Košenina.

19. See, for example, Spieß, "Fritzel mit dem roten Mützel," in *Biographien der Wahnsinnigen*, 77–78: "Könnten Sie in mein Herz blicken, so würden Sie finden, daß dies mein einziger, mein sehnlichster Wunsch ist."

20. "Mir kam es immer als eine Tücke, als ein Spionieren vor, wenn ich einen gegenwärtigen Menschen in seine Elemente zerlegen und seinen sittlichen Eigenschaften dadurch auf die Spur kommen wollte." Johann Wolfgang von Goethe, *Dichtung und Wahrheit*, ed. Klaus Detlef Müller, in Johann Wolfgang von Goethe, *Werke*, vol. 5 (Insel Verlag, 1998), 675.

21. On *eloquentia cordis*, see Ursula Geitner, *Die Sprache der Verstellung: Studien zum rhetorischen und anthropologischen Wissen im 17. und 18. Jahrhundert* (Niemeyer, 1992).

22. For a reassessment of this field of problems in relation to literary methodology and the active use of the notion of heteronomy, see Irene Albers, Marcus Hahn, and Frederic Ponten, eds., *Heteronomieästhetik der Moderne* (De Gruyter, 2022).

23. See Michael Neve, "Medicine and Literature," in *Companion Encyclopedia of the History of Medicine*, ed. William F. Bynum and Roy Porter, vol. 2 (Routledge, 1983), 1520–35, 1525; Sergio Moravia, "The Enlightenment and the Sciences of Man," *History of Science* 18 (1980): 247–68, 253; and Peter Gay, "Enlightenment as Medicine and Cure," in *The Age of the Enlightenment: Studies Presented to Theodore Bes-*

terman, ed. Theodore Besterman and W. H. Barber (Oliver & Boyd, 1967), 375–86, 375. "Master discourse" is a formulation in David B. Morris, "The Marquis De Sade and the Discourses of Pain: Literature and Medicine at the Revolution," in *The Languages of Psyche: Mind and Body in Enlightenment Thought*, ed. G. S. Rousseau (University of California Press, 1990), 291–330, 297.

24. See also G. J. Barker-Benfield, *The Culture of Sensibility: Sex and Society in Eighteenth-Century Britain* (University of Chicago Press, 1992), 15: "the flow of nerve ideas from scientists and diagnosticians into literature" refers to Cheyne, Sterne, and Rousseau among others.

25. On the century of Mechanism in Europe, see Anita Guerrini, "The Varieties of Mechanical Medicine: Borelli, Malpighi, Bellini and Pitcairne," in *Marcello Malpighi: Anatomist and Physician*, ed. Domenico Bertoloni Meli (Leo S. Olschki, 1997), 111–28; and Guido Giglioni, "The Machines of the Body and the Operations of the Soul in Marcello Malpighi's Anatomy," in ibid., 149–74.

26. See Theodore M. Brown, "From Mechanism to Vitalism in Eighteenth-Century English Physiology," *Journal of the History of Biology* 7, no. 2 (1974): 179–216, 179.

27. See Stephen Gaukroger, *The Collapse of Mechanism and the Rise of Sensibility: Science and the Shaping of Modernity, 1680–1760* (Oxford University Press, 2010).

28. Laurence Sterne, *The Life and Opinions of Tristram Shandy, Gentleman*, in *The Florida Edition of the Works of Laurence Sterne*, ed. Melvyn New and Joan New, vol. 1, *The Text* (University Presses of Florida, 1978), 82–84. The punctuation reproduced herehas been adjusted.

29. Sterne, *Tristram Shandy*, 1:85.

30. Sterne, *Tristram Shandy*, 1:83.

31. See my introduction.

32. On the paradoxical, blinding effect of this transparency, see Elena Fabietti, "'A Dioptrical Beehive': Die Repräsentation von Innerlichkeit zwischen Sichtbarkeit und Verblendung in Laurence Sternes *Tristram Shandy*," in *Alles Verblendung? Was wir nicht wahrnehmen können, sollen, wollen*, ed. Sebastian Donat et al. (Aisthesis, 2022), 51–59.

33. Kate E. Tunstall, "The Early Modern Embodied Mind and the Entomological Imaginary," in *Mind, Body, Motion, Matter: Eighteenth-Century British and French Literary Perspectives*, ed. Mary Helen McMurran and Alison Conway (University of Toronto Press, 2016), 202–29, 208.

34. A paradigmatic example is the work of Malpighi. On microscopy in Malpighi's anatomy, see Domenico Bertoloni Meli, *Mechanism, Experiment, Disease* (Johns Hopkins University Press, 2011), esp. 42.

35. Robert Hooke, *Microscopic Observations: Or, Dr. Hooke's Wonderful Discoveries by the Microscope, illustrated with thirty-three copperplates* (London: Robert Wilkinson, 1780), 48.

36. Bertoloni Meli, *Mechanism, Experiment, Disease*, 177.

37. Translated in Gaukroger, *Collapse of Mechanism*, 400. The original passage appears in Théophile de Bordeu, *Recherches Anatomiques sur la Position des Glandes et sur leur Action* (Paris: Chez G. F. Quillau, 1751), 452: "Nous comparons le corps vivant, pour bien sentir l'action particulière de chaque partie, à un essaim d'abeilles qui se ramassent en pelotons, et qui se suspendent à un arbre en manière de grappe . . . ; elle est un tout collé à un branche d'arbre, par l'action de bien des abeilles qui doivent agir ensemble, pour se bien tenir. . . . L'application est aisée: les organes du corps sont liés les unes avec les autres; ils ont chacun leur district et leur action; les rapports des ces actions, l'harmonie qui en résulte, font la santé."

38. On microscopy as an aesthetic enterprise, see Stafford, "Voyeur or Observer," 105–6: "Was microscopy a scientific enterprise or an aesthetic one? . . . In the absorbing illustrated narratives of Henry Baker, John Hill, and René-Antoine Ferchault de Réaumur, the diminutive picture functioned as an efficient image-storer for the violence of the macrocosmic world. Using the analogical method—supreme in the eighteenth century—these authors vied with their engravers in conveying the vivid impression that life down under the lens was even more bloody and rapacious than existence above it. Before Darwin, they showed nature to be read in miniature tooth and claw."

39. "Indeed, Tristram seems to hold that the operations of mind and body are largely mechanical. His father and uncle behave like a pair of clockwork toys, responding in programmed ways in conversation with each other." Judith Hawley, "The Anatomy of *Tristram Shandy*," in *Literature & Medicine During the Eighteenth Century*, ed. Marie Mulvey Roberts and Roy Porter (Routledge, 1993), 84–100, 85.

40. Judith Hawley, "*Tristram Shandy*, Learned Wit, and Enlightenment Knowledge," in *The Cambridge Companion to Laurence Sterne*, ed. Tom Keymer (Cambridge University Press, 2009), 34–48, esp. 45–46. Furthermore, considering Sterne through stark oppositional terms may be extremely misleading, as, indeed, his reception in Europe in the twentieth century showed, with Sterne labeled as a "sentimentalist" (mostly thanks to his last work, *A Sentimental Journey*) and his aesthetics ascribed to sensibility. For a discussion of this misrepresentation, see Jens Martin Gurr, *Tristram Shandy and the Dialectic of Enlightenment* (C. Winter, 1999), 150–55.

41. For example, in a famous passage about Shandeism: "True *Shandeism* . . . opens the heart and lungs, and like all those affections which partake of its nature, it forces the blood and other vital fluids of the body to run freely thro' its channels, and makes the wheel of life run long and chearfully [*sic*] round" (Sterne, *Tristram Shandy*, 1:401), modeled on James Mackenzie's *History of Health, and the Art of Preserving It* (1758). The example is given in Hawley, "*Tristram Shandy*, Learned Wit," 42. For more details on the medical authors quoted by Sterne, see Hawley, "*Tristram Shandy*, Learned Wit."

42. "Generation and fetal development, anatomy and surgery, and practical medicine in the form of diet, regimen, and the rival merits of hot spirits and cold bathing, all find their place here" (Hawley, "*Tristram Shandy*, Learned Wit," 35).

Famous iatrochemists were Archibald Pitcairne (1652–1713), Richard Mead (1673–1754), and Herman Boerhaave (1668–1738). I will refer to some of these figures later in this chapter. Hawley underlines that "in the eighteenth century, Iatromechanists and Vitalists were replacing the old system of humors with new anatomies which could be exploited by novelists to describe and account for the feelings of their characters" (Hawley, "Anatomy of *Tristram Shandy*," 85).

43. Sterne, *Tristram Shandy*, 1:85.

44. Juliet McMaster, "'Uncrystalized Flesh and Blood': The Body in *Tristram Shandy*," in *Laurence Sterne*, ed. Marcus Walsh (Longman, 2002), 95–111, 97; Hawley, "Anatomy of *Tristram Shandy*," 85. As Katherine Kickel has written: "Sterne's subject is man rather than machine" (Katherine E. Kickel, *Novel Notions: Medical Discourse and the Mapping of the Imagination in Eighteenth-Century English Fiction* (Routledge, 2007), 91.

45. McMaster "'Uncrystalized Flesh and Blood,'" 105.

46. Roy Porter, "The Whole Secret of Health: Mind, Body and Medicine in *Tristram Shandy*," in *Nature Transfigured: Science and Literature, 1700–1900*, ed. John Christie and Sally Shuttleworth (Manchester University Press, 1989), 61–84, 69.

47. Laurence Sterne, *The Letters, Part 1: 1739–1764*, *The Florida Edition of the Works of Laurence Sterne*, vol. 7, ed. Melvyn New and Joan New (University Presses of Florida, 2009), 80.

48. Gurr, *Tristram Shandy and the Dialectic of Enlightenment*, 74.

49. Sterne, *Tristram Shandy*, 1:350.

50. See Mirko D. Grmek, "A Survey of the Mechanical Interpretations of Life from the Greek Atomists to the Followers of Descartes," in *Biology, History, and Natural Philosophy*, ed. Allen P. Breck and Wolfgang Yourgrau (Plenum, 1974), 181–95.

51. Domenico Bertoloni Meli, "Machines and the Body Between Anatomy and Pathology," in *L'automate: Modèle métaphore machine merveille*, ed. Aurélia Gaillard (Presses universitaires de Bordeaux, 2013), 53–68, 53.

52. The work was published only twelve years after Descartes's death in a Latin translation, *De homine*. The French version, *L'Homme*, appeared in 1664. See René Descartes, *Treatise of Man*, ed. and trans. Thomas Steele Hall (Harvard University Press, 1972) (on *L'Homme*, see Hall's introductory notes, xxiv). In *Discours de la Méthode*, Descartes had, however, paraphrased part of the unpublished work.

53. Descartes, *Treatise of Man*, 2–4.

54. Descartes's mechanistic view of the human being was rooted in a corpuscular conception of the physical world. Crucial to Descartes's contribution to the debate over the functions of the soul is his refusal of the multiple functions of the soul (the origins of which are Aristotelian; see chap. 1). For Descartes, the proper soul is exclusively the rational one, and it is not the cause of the life of the body (see the commentary by Thomas Steel Hall in Descartes, *Treatise of Man*, 113–15 and n158). At the same time, *The Treatise of Man* lays the foundations of the interaction between body and soul in the brain itself (ibid., xxxi).

55. See Roger French, "The Anatomical Tradition," in Bynum and Porter, *Companion Encyclopedia of the History of Medicine*, 81–101, 95; and Guerrini, "Varieties of Mechanical Medicine," 112.

56. Marcello Malpighi, "Risposta del Dottor Marcello Malpighi alla Lettera Intitolata 'de recentiorum medicorum studio dissertatio epistolaris ad amicum,'" in *Opere Scelte*, ed. Luigi Belloni (UTET, 1967), 497–631, 512 (my translation). The passage is located in a text that is formatted as a quarrelsome answer to a colleague in Bologna, Domenico Sbaraglia, who claimed, against a "rational medicine," that medicine had to be an exclusively empirical science. The polemics has to be understood in terms of mechanism versus empiricism. For Malpighi, "empirical medicine is a heresy or scisma" ("Risposta," 511). Be that as it may, this antiempirical stand didn't translate completely into medical practice. Malpighi, in his clinical activity, still employed many of the traditional humoral medicine remedies (Guerrini, "Varieties of Mechanical Medicine," 118). This split between the praxis and the theory of medicine may be understood as a consequence of the fact that anatomy belonged to the field of natural philosophy more than to that of medicine: anatomy was, in fact, not yet mandatory training for aspiring physicians (Guerrini, 119). More in general, medical practice and research careers were clearly distinct routes (Guerrini, 128). On Sbaraglia and Malpighi, see also Bertoloni Meli, "Machines and the Body," 61–62.

57. Malpighi, "Risposta," 513–14. On the meaning of *a priori* for Malpighi, see Bertoloni Meli, "Machines and the Body," 66, who explains that *a priori* implied, for Malpighi, that physiological causes "could be investigated by mechanical devices."

58. Malpighi, "Risposta," 516–17. Where a model-machine cannot always reproduce the actual phenomenon (because the latter often occurs at the microscopic level), "the intellect, trained in mechanics, can get there through reasoning" (514).

59. Giovanni A. Borelli, *On the Movement of Animals* (Springer, 1989), 60, 399; the work *De Motu Animalium* was written by 1675 but published posthumously in 1680–81. It is worth noticing that Borelli was critical of Descartes because he deemed Descartes's machines to be arbitrary and not mathematically deduced, unlike the ones he wanted to describe (Guerrini, "Varieties of Mechanical Medicine," 114). Though not a physician himself, Borelli wrote *Delle cagioni delle febbri maligne della Sicilia negli anni 1647 e 1648* (1649), a work that had quickly become an iatromechanical manifesto (Guerrini, 113). It is important to note here that the practice of iatromechanism presented consistent variations, both geographically and individually, from one doctor or anatomist to the other. As Brown has argued, the London College of Physicians, which had accepted iatromechanism by 1700, advocated for a very different kind of Mechanism than its Italian counterpart. See Theodore M. Brown, "The College of Physicians and the Acceptance of Iatromechanism in England, 1665–1695," *Bulletin of the History of Medicine* 44, no. 1 (1970): 12–30. Mainly, anatomy was not the basis of English Mechanism, as it was considered too empirical a foundation for medical science (Guerrini, 124). The leading

figure of Mechanism in England, Archibald Pitcairne (1652–1713), was not even trained in anatomy and proposed mathematics instead of anatomy as the foundation of mechanistic medicine (Guerrini, 124). In other words, simply hearkening back to the common denominator of "Mechanism" does not seal the gaps between motivations and aims that sustain different mechanistic theories and practices. Not only did the most strenuous mechanists often continue to apply traditional "empiricist" medicine in their therapeutic practice (the speculative dimension of Mechanism and the practice of medicine remained two parallel practical and social dimensions), but different schools of Mechanism emphasized in some cases the merely theoretical representation of the body as governed by mechanics and in other cases the concrete, mechanic structures of human anatomy.

60. Giorgio Baglivi, *Opera omnia medico-pratica et anatomica* (Lipsia: Kühn, 1727), 155 (my translation).

61. Julien Offray de La Mettrie, *Machine Man and Other Writings*, ed. and trans. Ann Thomson (Cambridge University Press, 1996), 34; Julien Offray de La Mettrie, *L'Homme machine*, in *L'Homme machine: A Study in the Origins of an Idea*, ed. Aram Vartanian (Princeton University Press, 1960), 190.

62. La Mettrie, *Machine Man*, 34; La Mettrie, *L'Homme machine*, 191.

63. On La Mettrie, see Kathleen Wellman, *La Mettrie: Medicine, Philosophy, and Enlightenment* (Duke University Press, 1992). Wellman's reading of La Mettrie claims for his work the thesis that "man is no exception to the uniformity [of] nature" (181). Wellman insists on La Mettrie's irregularity and his misreading in historiography, where he has been seen as a follower of Descartes: a misunderstanding based on the positive remarks he makes on Descartes in *L'Homme machine*, which should be however understood ironically (183). La Mettrie had been a student of Hermann Boerhaave, probably the most famous physician of the century and a real cult figure for generations of medical students trained in his headquarters in Leiden (82–83).

64. La Mettrie, *Machine Man*, 39; La Mettrie, *L'Homme machine*, 197.

65. George Makari gives an enticing account of La Mettrie's work within the frame of the historical development of Vitalism in George Makari, *Soul Machine: The Invention of the Modern Mind* (Norton, 2015), 244: "Man was a weird, organic, alive kind of machine. Thus, older conceptions of mechanics needed to be revised into more dynamic models to explain human beings. Man was a 'machine that winds its own springs.'"

66. Malpighi mentions the *Statua* in the polemical text from which I quoted earlier: "Risposta," 514 ("il Reisellio [*sic*] ne ha composto una macchina colla quale mostra tutte le suddette operazioni e fenomeni").

67. The entire publication's title was *Miscellanea curiosa sive ephemeridum medico-physicarum Germanicarum Academiae Caesareo-Leopoldinae Naturae Curiosorum*. In 1687, the Academy was called Leopoldina after Leopold I, who had officially recognized it in 1677. For a more detailed account of Reisel's biography, see Ralf Bröer, *Salomon Reisel (1625–1701): Barocke Naturforschung eines Leibarztes im*

Banne der Mechanistischen Philosophie (Deutsche Akademie der Naturforscher Leopoldina, 1996).

68. Salomon Reisel, "Statua humana circulatoria," *Miscellanea curiosa: Sive ephemeridum medico-physicarum Germanicarum*, anni IX (1680): 1–22, 8 (translation from the Latin is mine).

69. See also Bröer, *Salomon Reisel*, 57; and Mattia Mantovani, "Descartes' Man Under Construction: The Circulatory Statue of Salomon Reisel, 1680," *Early Science and Medicine* 25 (2020): 101–34.

70. Reisel, "Statua humana circulatoria," 20. The syphons were working continuously, provided that a fluid was regularly replenished (in other words, a pump was not necessary).

71. The statue's heart was planned to be made of glass but ultimately was rendered in leather: Mantovani, "Descartes' Man," 122.

72. On the pineal gland, see Descartes, *Treatise of Man*, 86, esp. note 135.

73. See Bertoloni Meli, "Machines and the Body," 68.

74. "Reisel was not aiming to mimic the outward appearance of animal motions, but wanted to re-enact the very life processes, so as to demonstrate their mechanical nature." Mantovani, "Descartes' Man," 133.

75. Georges Canguilhem, *Knowledge of Life*, ed. Paola Marrati and Todd Meyers (Fordham University Press, 2008), 80.

76. Canguilhem, *Knowledge of Life*, 78; Derek J. de Solla Price, "Automata and the Origins of Mechanism and Mechanistic Philosophy," *Technology and Culture* 5, no. 1 (1964): 9–23, 9.

77. Bröer, *Salomon Reisel*, 54. Reisel himself attests to this in a later appendix to his contribution, "Sipho Würtembergicus, sive Sipho inversus cruribus æquialtis fluens & refluens hactetus inauditus," *Miscellanea curiosa: Sive ephemeridum medico-physicarum Germanicarum* (1684), 461–472, 464.

78. On Vaucanson and Reisel, see Mantovani, "Descartes' Man," 132. On Vaucanson and the Swiss clockmakers Pierre-Jacquet Drosz and his son Henri-Louis, see Silvio A. Bedini, "The Role of Automata in the History of Technology," *Technology and Culture* 5, no. 1 (1964): 24–42, esp. 31. On the history of android automata, which began with Hans Bullmann in Nuremberg in the sixteenth century, and clockmaking tradition in Nuremberg and Augsburg, as well as on the legendary rumor that Descartes himself in his youth had imagined/envisioned and perhaps managed to build an automaton, see de Solla Price, "Automata," 22–23.

79. See Moravia, "Enlightenment and the Sciences." The controversy has been problematized as metatheoretical rather than actual; see Hilde Hein, "The Endurance of the Mechanism-Vitalism Controversy," *Journal of the History of Biology* 5, no. 1 (Spring 1972): 159–88.

80. Moravia, "Enlightenment and the Sciences," 252–53; Colm Kiernan, *The Enlightenment and Science in Eighteenth-Century France* (Voltaire Foundation, 1973), 13–14.

81. Theodore M. Brown, "From Mechanism to Vitalism," 188; Sergio Moravia,

"From *Homme Machine* to *Homme Sensible*: Changing Eighteenth-Century Models of Man's Image," *Journal of the History of Ideas* 39, no. 1 (1978): 45–60, 49.

82. Philippe Huneman and Charles T. Wolfe, "Man-Machines and Embodiment: From Cartesian Physiology to Claude Bernard's 'Living Machine,'" in *Embodiment: A History*, ed. Justin E. H. Smith (Oxford University Press, 2017), 241–76, 256. A similar analysis of this historical debate in relation to the emergence of Montpellier Vitalism that we are about to see is also in Charles T. Wolfe, "Models of Organic Organization in Montpellier Vitalism," *Early Science and Medicine* 22, nos. 2–3 (2017): 229–52 (see esp. 236).

83. Jacques Roger, *Les Sciences de la vie dans la pensée française du XVIIIe siècle: La Génération des animaux de Descartes à l'encyclopédie* (Paris: Colin, 1963), 164.

84. Canguilhem, *Knowledge of Life*, 90. For a concrete example of the limits of mechanistic physiology, see Brown, "From Mechanism to Vitalism," 186–87; 194.

85. Moravia, "Enlightenment and the Sciences," 250.

86. For a detailed and comprehensive account of all the different contributions to Montpellier Vitalism, see Elizabeth A. Williams, *A Cultural History of Medical Vitalism in Enlightenment Montpellier* (Ashgate, 2003), especially on the beginnings of the movement (147). Williams shows, on the one hand, the entanglements of these figures with forms of Protestantism and, on the other, their wish to overcome classic Cartesianism and its predicaments (83). On Montpellier Vitalism and its relation to the long and somehow hazy history of Vitalisms, see Wolfe, "Models of Organic Organization."

Although in this chapter I deal with Montpellier Vitalism, another important European center for Vitalist thought had been, since at least the 1720s, the northern town of Halle in Germany. Here the figure of Georg-Ernst Stahl had been challenging Cartesianism and its mind-body dualism in name of a "holistic organism" directed by a soul (*anima*), whence the label of "animism" assigned to Stahlian doctrines.

87. For a lively narrative reconstructing the scientific and biographic adventures between Paris *philosophes* and Montpellier doctors, see Makari, *Soul Machine*, 245–66. On the reasons for the end of the collaboration with the *Encyclopédie*, see Williams, *A Cultural History*, 147–48.

88. Théophile de Bordeu's entry "Crise" (which concerned the already Hippocratic issue of the critical days in the course of a disease) and Venel's article "Chymie," in particular, together came to be seen as "a kind of vitalist manifesto" (Williams, *A Cultural History*, 161).

89. Williams, *A Cultural History*, 224.

90. First, as we have seen, these narratives are interlocked, as Wolfe has shown in his "Man-Machines and Embodiment" and "Models of Organic Organization." More in general, to avoid sweeping historiographical claims and oversimplified narratives, it is wise to rely on the methodological suggestion by Lorraine Daston and Peter Galison in *Objectivity*: "Instead of the analogy of a succession of political regimes or scientific theories, each triumphing on the ruins of its predecessor, imag-

ine new stars winking into existence, not replacing old ones but changing the geography of the heavens" (Lorraine Daston and Peter Galison, *Objectivity* [Zone, 2007], 18).

91. Anne C. Vila, *Enlightenment and Pathology: Sensibility in the Literature and Medicine of Eighteenth-Century France* (Johns Hopkins University Press, 1998), 16; Moravia, "From *Homme Machine* to *Homme Sensible*," 58; Marco Menin, "Jean-Jacques Rousseau vitaliste: La Moralisation de l'hygiène médicale entre régime diététique et éthique alimentaire," *Nuncius* 27 (2012): 81–109, 105.

92. Original is in Denis Diderot and Jean Le Rond d'Alembert, *Encyclopédie, ou dictionnaire raisonné des sciences, des arts et des métiers*, vol. 14 (Neuchatel: Chez Samuel Faulche, 1765), 937 (my emphasis). The passage is also quoted in part in Vila, *Enlightenment and Pathology*, 53; here I have integrated the rest of it by my translation.

93. The idea of such a missing link is in Makari, *Soul Machine*, 257, where he insists on the crucial function of sensibility in overcoming both Mechanism and Animism: "Theories of sensibility opened the door for an embodied mind. Perhaps man did not reduce down to a ticking contraption, but rather the very idea of the body must be expanded to possibly include complex, dynamic functions, including wanting, thinking, and choosing" (257–58).

94. In the words of Stephen Gaukroger: "a distinctive feature of mid-eighteenth century thought is the way in which questions of cognition, morality, and civic responsibilities come to be grounded in sensibility" (Gaukroger, *Collapse of Mechanism*, 402).

95. For a broader overview of the intellectual history of the notion, see Henry Martin Lloyd, ed., *The Discourse of Sensibility: The Knowing Body in the Enlightenment* (Springer, 2013).

96. Lloyd, *The Discourse of Sensibility*, 4.

97. Sensibility could so become "the term generally accepted by eighteenth-century thinkers to designate this all-encompassing vital property" (Vila, *Enlightenment and Pathology*, 15–16).

98. Vila, *Enlightenment and Pathology*, 20.

99. Albrecht von Haller, *A Dissertation on the Sensible and Irritable Parts of Animals*, trans. M. Tissot (London: J. Nourse, 1755), 4.

100. Vila, *Enlightenment and Pathology*, 25.

101. See Philippe Huneman, "Montpellier Vitalism and the Emergence of Alienism in France (1750–1800): The Case of the Passions," *Science in Context* 21, no. 4 (2008): 615–47, 617.

102. See chap. 2n37.

103. Denis Diderot, *Rameau's Nephew and D'Alembert's Dream*, ed. and trans. Leonard Tancock (Penguin Classics, 2011), 170; Diderot, *Le Rêve de d'Alembert*, in *Œuvres Philosophiques*, ed. Paul Vernière (Garnier Frères, 1964), 293: "Je puis donc assurer a présent à toute la terre qu'il n'y a aucune différence entre un médecin qui veille et un philosophe qui rêve."

104. Diderot, *D'Alembert's Dream*, 167; Diderot, *Le Rêve de d'Alembert*, 289.

105. Diderot, *D'Alembert's Dream*, 219; Diderot, *Le Rêve de d'Alembert*, 367.

106. See, in particular, the first "Entretien" between Diderot and d'Alembert (Diderot, *Entretien*, in *Œuvres Philosophiques*, 275–76). Also: "Il n'y a plus qu'une substance dans l'univers, dans l'homme, dans l'animal" (278).

107. Diderot, *D'Alembert's Dream*, 188. Diderot, *Le Rêve de d'Alembert*, 322: "Si cette infinie diversité de touchers n'existait pas, on saurait qu'on éprouve du plaisir ou de la douleur, mais on ne saurait où les rapporter. Il faudrait le secours de la vue. Ce ne serait plus une affaire de sensation, ce serait une affaire d'expérience et d'observation."

108. The model is still present to some extent in *Le Rêve de d'Alembert*, where it is voiced by the pragmatic Mademoiselle d'Espinasse.

109. Diderot and D'Alembert, *Encyclopédie*, vol. 11, 299 (my translation; emphasis in the original).

110. Dominique Boury, "Irritability and Sensibility: Key Concepts in Assessing the Medical Doctrines of Haller and Bordeu," *Science in Context* 21, no. 4 (2008): 521–35, 528.

111. In the text my translation of Diderot and D'Alembert, *Encyclopédie*, vol. 15, 38:

> SENSIBILITÉ, SENTIMENT, (*Médecine*) la faculté de sentir, le principe sensitif, ou le sentiment même des parties, la base & l'agent conservateur de la vie, l'animalité par excellence, le plus beau, le plus singulier phénomène de la nature, &c.
>
> La sensibilité est dans le corps vivant, une propriété qu'ont certaines parties de percevoir les impressions des objets externes, & de produire en conséquence des mouvements proportionnés au degré d'intensité de cette perception.

112. In the text my translation of Louis de Jaucourt, *Encyclopédie*, vol. 15, 52:

> SENSIBILITÉ, (Morale.) disposition tendre & délicate de l'ame [*sic*], qui la rend facile à être émue, à être touchée. La sensibilité d'ame, dit très bien l'auteur des mœurs, donne une sorte de sagacité sur les choses honnêtes, & va plus loin que la pénétration de l'esprit seul. Les ames sensibles peuvent par vivacité tomber dans des fautes que les hommes à procédés ne commettraient pas; mais elles l'emportent de beaucoup par la quantité des biens qu'elles produisent. Les ames sensibles ont plus d'existence que les autres: les biens & les maux se multiplient à leur égard. La réflexion peut faire l'homme de probité; mais la sensibilité fait l'homme vertueux. La sensibilité est la mère de l'humanité, de la générosité; elle sert le mérite, secourt l'esprit, & entraîne la persuasion à sa suite.

113. From a different but related perspective, on sensibility as a gender-marker, see Barker-Benfield, *The Culture of Sensibility*.

114. Diderot, *D'Alembert's Dream*, 212; Diderot, *Le Rêve de d'Alembert*, 356–57: "Mais qu'est-ce qu'un être sensible? Un être abandonné à la discrétion du dia-

phragme. Un mot touchant a-t-il frappé l'oreille? Un phénomène singulier a-t-il frappé l'œil? Et voilà tout à coup le tumulte intérieur qui s'élève, tous les brins du faisceau qui s'agitent, le frisson qui se répand, l'horreur qui saisit, les larmes qui coulent, les soupirs qui suffoquent, la voix qui s'interrompt, l'origine du faisceau qui ne sait ce qu'il devient; plus de sang froid, plus de raison, plus de jugement, plus d'instinct, plus de ressource."

115. This critical gesture is rather typical of literature imbricated with the notion of sensibility. In Sterne's *Tristram Shandy*, sensibility is exposed to subtle irony (Thomas Keymer, "*A Sentimental Journey* and the Failure of Feeling," in *The Cambridge Companion to Laurence Sterne*, ed. Tom Keymer, 89–90). The case of Sterne is particularly telling, since his fame was long linked to his work *A Sentimental Journey to France and Italy*, which more or less ironically designated him a writer of "sensibility."

116. A slightly different, though resonating, genealogy for the pervasive notion of sensibility than the one I have outlined here, following Anne Vila's emphasis on Haller's physiology, is that proposed by the historian of medicine Georges Rousseau in Georges Rousseau, "Nerves, Spirits, and Fibres: Toward Defining the Origins of Sensibility," in *Nervous Acts: Essays on Literature, Culture, and Sensibility*, ed. G. S. Rousseau (Palgrave Macmillan, 2004), 157–84. For Georges Rousseau, the "cults" of sensibility that exploded in the second half of the eighteenth century and that produced a proper genre of novels owed their aesthetics not so much to the contemporary scientific debates about sensibility as to the previous ones concerning the brain and the nerves. The claim is that sensibility was already a product of the late seventeenth century (especially of the medical work of Thomas Willis, hence of the philosophy of his pupil John Locke) and is tied to the discovery of the nerves as vectors of the brain's impulses (Willis), before expanding into a broader category defining at the same time perception and morals. For Georges Rousseau, the physiology of the nerves yielded the aesthetics of sensibility through a sequence of connections: "the greater one's nervous sensibility, the more one is capable of delicate writing" (171).

117. Jean-Jacques Rousseau, *Confessions*, trans. Angela Scholar, ed. Patrick Coleman (Oxford University Press, 2001), 5. "Je forme une entreprise qui n'eut jamais d'exemple et dont l'exécution n'aura point d'imitateur. Je veux montrer à mes semblables un homme dans toute la vérité de la nature; et cet homme ce sera moi." Jean-Jacques Rousseau, *Les Confessions*, in *Œuvres complètes*, vols. 1–4, ed. Bernard Gagnebin and Marcel Raymond (Gallimard, 1959–95), 1:5. Quotations are taken from this edition, with spelling occasionally modernized for clarity.

118. Rousseau, *Les Confessions*, 5. "Moi seul. Je sens mon cœur et je connais les hommes. Je ne suis fait comme aucun de ceux que j'ai vus; j'ose croire n'être fait comme aucun de ceux qui existent. Si je ne vaux pas mieux, au moins je suis autre. Si la nature a bien ou mal fait de briser le moule dans lequel elle m'a jeté, c'est ce dont on ne peut juger qu'après m'avoir lu" (*Les Confessions*, 5).

119. "Un ouvrage unique et utile, lequel peut servir de première pièce de com-

paraison pour l'étude de l'homme, qui certainement est encore à commencer" (Rousseau, *Les Confessions*, 3). The text of this note, published in the French edition cited, was retrieved from a note in the Geneva manuscript of the *Confessions*. It is not reproduced in the English translations. See the "Notes et variantes" to *Les Confessions*, 1230.

120. Rousseau, *Confessions*, 7. "De tous les dons que le Ciel leur avait départis, un *cœur sensible* est le seul qu'ils me laissèrent; mais il avait fait leur bonheur, et fit tous les malheurs de ma vie" (Rousseau, *Les Confessions*, 7; my emphasis).

121. Vila, *Enlightenment and Pathology*, 182–84. According to Vila, by acknowledging sensibility as the main source of human action, Rousseau's ethical recipe consisted in controlling sensibility's powers. The relation of Rousseau with medicine is complex, however, as argued in Menin, "Jean-Jacques Rousseau vitaliste," 95–96. Rousseau saw medicine as fundamentally dangerous, except for medicine intended as hygiene. In *Émile*, he writes: "la seule partie utile de la médecine est l'hygiène; encore l'hygiène est-elle moins une science qu'une vertu" (J.-J. Rousseau, *Émile*, in *Œuvres complètes*, vol. 4, 271). For Menin, Rousseau establishes a link between the physical and the moral plane and thereby gives his main contribution to the discourse of Vitalism (108). From a biographical standpoint, while consistently critical of medicine in his work, Rousseau was in close exchange with his compatriot, the Swiss doctor André David Tissot (see Vila, *Enlightenment and Pathology*, 188).

122. Rousseau, *Confessions*, 8; Rousseau *Les Confessions*, 8.

123. "Exister pour nous, c'est sentir; notre sensibilité est incontestablement antérieure à notre intelligence" (Rousseau, *Émile*, 600). To further retrieve the sources of Rousseau's notion of sensibility, it would be productive to look at the theory of sensations elaborated by Etienne Bonnot de Condillac (1715–80), whose person and work Rousseau was familiar with. The description of the statue that starts living thanks to sensations in *Traité des Sensations* (1754), especially, would lead to a productive comparison of the notions. Etienne de Condillac, *Traité des Sensations; Traité des Animaux* (Fayard, 1984), 9, 235. For an investigation of this genealogy, see Udo Thiel, "Self and Sensibility: From Locke to Condillac and Rousseau," *Intellectual History Review* 25, no. 3 (2015): 257–77, 265; and Makari, *Soul Machine*, 277. On Rousseau and Condillac, see also "Notes et variantes," in Rousseau, *Œuvres complètes*, vol. 1, 1372.

124. Rousseau, *Confessions*, 110.

> Deux choses presque inalliables s'unissent en moi sans que j'en puisse concevoir la manière: un tempérament très ardent, des passions vives, impétueuses, et des idées lentes à naître, embarrassées, et qui ne se présentent jamais qu'après coup. On dirait que mon cœur et mon esprit n'appartiennent pas au même individu. Le sentiment plus prompt que l'éclair vient remplir mon âme, mais au lieu de m'éclairer il me brûle et m'éblouit. *Je sens tout et je ne vois rien.* Je suis emporté mais stupide; il faut que je sois de sang-froid pour penser. Ce qu'il y a d'étonnant est que j'ai cependant le tact assez sûr, de la pénétration, de la finesse même

pourvu qu'on m'attende: je fais d'excellents impromptus à loisir; mais sur le temps je n'ai jamais rien fait ni dit qui vaille. (Rousseau, *Les Confessions*, 113 [my emphasis])

125. Rousseau, *Confessions*, 40. "Un cœur trop affectueux, trop aimant, trop tendre, qui, faute d'en trouver d'existans [*sic*] qui lui ressemblent est forcé de s'alimenter de fictions" (Rousseau, *Les Confessions*, 41).

126. See Robert Erickson, *The Language of the Heart, 1600–1750* (University of Pennsylvania Press, 1997). For Erickson, the heart of early modern Europe was thick in significations, having accumulated a history of contradictory or overlapping anatomical functions.

127. Ernst Robert Curtius, *European Literature and the Latin Middle Ages* (Princeton University Press, 2013), 135–37.

128. Geitner, *Die Sprache der Verstellung*, 198.

129. Erickson, *The Language of the Heart*. Like Hippocrates, Aristotle located the sentient soul in the heart; Galen will relocate it in the brain (3–4). Meanwhile, the Bible assigned it both a sentient and thinking function (11–1w3).

130. "Posons pour maxime incontestable que les premiers mouvements de la nature sont toujours droits: il n'y a point de perversité originelle dans le cœur humain; il ne s'y trouve pas un seul vice dont on ne puisse dire comment et par où il y est entré" (Rousseau, *Émile*, 322).

131. Rousseau, *Rousseau juge de Jean-Jacques, Dialogues*, in *Œuvres complètes*, vol. 1, 860 (my translation).

132. Rousseau, *Discourse on the Science and the Arts*, in Jean-Jacques Rousseau, *Basic Political Writings*, trans. and ed. Donald A. Cress (Hackett, 2011), 7. "La nature humaine, au fond, n'étoit pas meilleure; mais les hommes trouvaient leur sécurité dans la facilité de se pénétrer réciproquement" (Rousseau, *Discours sur les sciences et les arts*, in *Œuvres complètes*, vol. 3, 8).

133. Jean Starobinski, *Jean-Jacques Rousseau, Transparency and Obstruction* (University of Chicago Press, 1988), 15, 18, 24.

134. Quoted in Starobinski, *Transparency and Obstruction*, 227.

135. Quoted in Starobinski, *Transparency and Obstruction*, 256. On the unfinished manuscript of the *Institutions chimiques* and Rousseau's toying with the idea of glass and of vitreous humans, see "Notes et variants," in Rousseau, *Œuvres complètes*, vol. 1, 1414.

136. Starobinski, *Transparency and Obstruction*, 257.

137. See Haller, *A Dissertation*, 6: "I took living animals of different kinds, and different ages, and after laying bare that part which I wanted to examine, I waited till the animal ceased to struggle or complain; after which I irritated the part, by blowing, heat, spirit of wine, the scalpel, lapis infinalis, oil of vitriol, and butter of antimony. I examined attentively, whether upon touching, cutting, burning, or lacerating the part, the animal seemed disquieted, made a noise, struggled, or pulled back the wounded limb, if the part was convulse, or if nothing of all this happened."

138. Rousseau, *Confessions*, 58.

> Dans l'entreprise que j'ai faite de me montrer tout entier au public, il faut que rien de moi ne lui reste obscur ou caché; il faut que je me tienne incessamment sous ses yeux; qu'il me suive dans tous les égarements de mon cœur, dans tous les recoins de ma vie; qu'il ne me perde pas de vue un seul instant, de peur que, trouvant dans mon récit la moindre lacune, le moindre vide, et se demandant: Qu'a-t-il fait durant ce temps-là, il ne m'accuse de n'avoir pas voulu tout dire. Je donne assez de prise à la malignité des hommes par mes récits, sans lui en donner encore par mon silence. (Rousseau, *Les Confessions*, 59–60)

139. On the rhetorical complex of transparency, visibility, and readability, see the classic analysis in Paul de Man, *Allegories of Reading: Figural Language in Rousseau, Nietzsche, Rilke, and Proust* (Yale University Press, 1979), 192–93.

140. Rousseau, *Confessions*, 110.

141. Rousseau, *Confessions*, 436. "Son cœur transparent comme le cristal ne peut rien cacher de ce qui s'y passe; chaque mouvement qu'il éprouve transmet à ses yeux et sur son visage" (Rousseau, *Les Confessions*, 860).

142. See Barbara Maria Stafford, *Body Criticism: Imaging the Unseen in Enlightenment Art and Medicine* (MIT Press, 1991), 85.

143. Marin Cureau de La Chambre, *L'Art de connoistre les hommes* (Amsterdam: Jacques le Jeune, 1660), 1–2 (my translation and my emphasis). Here is my (modernized) version of the French text:

> Celui-là n'avait pas raison, qui se plaignait autrefois, de ce que la nature n'avait pas mis une fenêtre au devant du Cœur, pour voir les pensées et les desseins des hommes. Non seulement parce que ce sont des choses qui ne tombent pas sous les sens, et que quand les yeux verraient tous le fonds et tous les replis du cœur, ils n'y pourraient rien remarquer qui leur en donnât la moindre connaissance. Mais encore parce que la Nature a pourvue à cette découverte, et a trouvé des moyens plus certains pour la faire, que n'eut été cette étrange ouverture que Momus s'était imaginé.

144. See the digression on dissimulation in Rousseau, *Rousseau, juge de Jean-Jacques*, 861.

145. For a discussion conducted between the field of rhetorics and the methodology of media studies in the German context, see Geitner, *Die Sprache der Verstellung*; Alexander Košenina, *Anthropologie und Schauspielkunst* (Niemeyer, 1995); and Koschorke, *Körperströme und Schriftverkehr*.

146. For a lively review of Rousseau's social misadventures, see Makari, *Soul Machine*, 275–97; and for the English misadventures of Rousseau in particular, see David Edmonds and John Eidinow, *Rousseau's Dog: Two Great Thinkers at War in the Age of Enlightenment* (Harper Collins, 2006).

147. As he announced in the *Préambule* to the *Confessions* (Rousseau, *Les Confessions*, 3).

Chapter 3

1. Later in this chapter, we will see how Paracelsus's description of the Homunculus assigns him a transparency of first degree—that is, a bodily transparency, the implications of which are, however, different from that conferred by the glass vial.

2. I use the generic indication of "tragedy" employed by Goethe himself, with the understanding that Goethe was aware of the irregular nature of his brand of tragedy. On Goethe and tragedy, see Klaudia Hilgers, *Entelechie, Monade und Metamorphose: Formen der Vervollkommnung im Werk Goethes* (Wilhelm Fink, 2002), 223–24. On the dramatic function of Homunculus in relation to other main characters, see Joachim Müller, *Die Figur des Homunculus in Goethes "Faust"* (Akademie-Verlag, 1963); and Katharina Mommsen, *Natur- und Fabelreich in Faust II* (De Gruyter, 1968).

3. Among the main psychological interpretations, see Carl Gustav Jung, *Psychologie und Alchemie* (Rascher, 1944); Karl Kerényi, *Das Ägäische Fest: Die Meergötterszene in Goethes Faust II* (Pantheon, 1941); Irene Gerber-Munch, "Goethe's Faust and the Myth of Modern Man," *Psychological Perspectives* 52, no. 2 (2009): 143–66; and Oskar Sahlberg, "The Alchemist's Love Story: The Creation of the Homunculus—Growth and Birth of the Self," in *Proceedings of the 12th International Conference on Literature and Psychoanalysis* (Lisbon, 1996) 65–68.

4. Manfred Osten, "Dr. Faust—ein Auslaufmodell der Evolution? Goethes Tragödie und die Verheißungen der Lebenswissenschaften," *Goethe-Jahrbuch* 124 (2007): 161–66.

5. Mary Baine Campbell, "Artificial Men: Alchemy, Transubstantiation, and the Homunculus," *Republics of Letters: A Journal for the Study of Knowledge, Politics, and the Arts* 1, no. 2 (2010): 4–15.

6. Bruce J. MacLennan, "Homunculus' Quest for a Body," in *Goethe's Faust and Cultural Memory*, ed. Lorna Fitzsimmons (Lehigh University Press, 2012), 178–215. The series of inspired readings goes on, as summarized in Rudolph Drux, "Homunculus oder Leben aus der Retorte: Zur Kulturgeschichte eines literarischen Motivs seit Goethe," in *Tales from the Laboratory: or, Homunculus Revisited*, ed. Rüdiger Görner (Iudicium, 2005), 91–104, 95. The spectrum of meanings of Homunculus in *Faust II* ranges from the contemporary chemical discovery of a synthesis of organic compounds to the parody of the contemporary literary scene, especially the idea of an ever-becoming *Universalpoesie* such as that propounded by Friedrich Schlegel (and never, for Goethe, actualized beyond theoretical statements). The latter reading has been advanced in Otto Höfler's controversial study, *Homunculus: Eine Satire auf A. W. Schlegel* (Böhlau, 1972).

7. I quote from Johann Wolfgang von Goethe, *Faust I & II*, trans. Stuart Pratt Atkins, ed. David E. Wellbery (Princeton University Press, 2014), 175. Goethe, *Faust: Der Tragödie zweiter Teil*, in *Dramatische Dichtungen*. In Johann Wolfgang von Goethe, *Werke. Hamburger Ausgabe in 14 Bänden*, vol. 3, ed. Erich Trunz (C. H. Beck, 1982). I will henceforth provide first the page number of the English text, then of the German edition. In the interest of the chapter's general economy,

quotations from the German text are limited to those instances in which they are essential to the argument. Bibliographical references to the German text are always supplied to allow interested readers to retrieve the original passages.

8. *Faust I & II*, 175; *Faust*, 209.

9. *Faust I & II*, 175; *Faust*, 209.

10. *Faust I & II*, 175; *Faust*, 210.

11. *Faust I & II*, 176; *Faust*, 210.

12. Lawrence M. Principe, *The Secrets of Alchemy* (University of Chicago Press, 2013), 86. On the material importance of glass for alchemy, and how product and process are historically interrelated, see Marco Beretta, *The Alchemy of Glass: Counterfeit, Imitation, and Transmutation in Ancient Glassmaking* (Science History Publications, 2009).

13. See Ronald D. Gray, *Goethe the Alchemist: A Study of Alchemical Symbolism in Goethe's Literary and Scientific Works* (Cambridge University Press, 1952), 102.

14. Principe, *The Secrets of Alchemy*, 124.

15. *Faust I & II*, 178; *Faust*, 212.

16. *Faust I & II*, 179; *Faust*, 214.

17. *Faust I & II*, 181; *Faust*, 216.

18. *Faust I & II*, 181; *Faust*, 216.

19. *Faust I & II*, 181; *Faust*, 216.

20. *Faust I & II*, 199; *Faust*, 238.

21. *Faust I & II*, 177; *Faust*, 211.

22. Johann Peter Eckermann, *Conversations of Goethe with Eckermann and Soret*, trans. John Oxenford (Cambridge University Press, 2011), 198. The original passage is in Johann Peter Eckermann, *Gespräche mit Goethe in den letzten Jahren seines Lebens*, in *Johann Wolfgang Goethe: Sämtliche Werke. Nach Epochen seines Schaffens. Münchner Ausgabe*, ed. Karl Richter, vol. 19 (Hanser Verlag, 1986), 339.

23. *Faust I & II*, 177; *Faust*, 212.

24. Joachim Müller has defined it as medial identification, in Müller, *Die Figur des Homunculus*, 17.

25. *Faust I & II*, 179; *Faust*, 214.

26. Eckermann, *Conversations*, 198; Eckermann, *Gespräche*, 339.

27. Eckermann, *Conversations*, 198; Eckermann, *Gespräche*, 339. Translation amended by me.

28. Eckermann, *Conversations*, 198; Eckermann, *Gespräche*, 339.

29. "I have doubted whether I ought not to put some verses into the mouth of Mephistopheles as he goes to Wagner, and the Homunculus is still in a state of formation, so that his co-operation may be expressed and rendered plain to the reader" (Eckermann, *Conversations*, 198; Eckermann, *Gespräche*, 339). On further, more radical, aspects of Mephistopheles's possible intervention in the generation of Homunculus, see Kathleen R. Snow, "Homunculus in Paracelsus, *Tristram Shandy*, and *Faust*," *Journal of English and Germanic Philology* 79, no. 1 (1980): 67–74, 71.

For further reflections on Homunculus as Mephistophelian creature, see Dieter Arendt, "Homunculus—der Geist 'in der Flasche' oder: Ein mephistophelisches Märchen," *Studi Germanici*, n.s., 45, no. 2 (2007): 289–300.

30. On the discovery and Goethe's interest, see Wolf von Engelhardt, *Goethe im Gespräch mit der Erde* (Metzler, 2003), 351. On the fascination of Goethe with the relation between inorganic and organic matter, see Wolfgang Hottner, *Kristallisationen. Ästhetik und Poetik des Anorganischen im späten 18. Jahrhundert* (Wallstein, 2020), esp. 244–246.

31. Gray, *Goethe the Alchemist*.

32. In light of Lawrence Principe's historiographical reassessment of alchemy (Principe, *The Secrets of Alchemy*), one can see how Gray's study incorrectly projects late nineteenth-century revivals of alchemy, inspired by distinct occultist fashions, on early modern alchemy (or *chymistry*, as Principe suggests calling it, in order to overcome the inaccurate disjunction of chemistry and alchemy for the early modern period). Gray, for example, often confuses magic with alchemy or attributes to alchemy the same quality of occultist investigations, not to mention his failure to credit the secretive recipes of alchemists as valid, whereas Principe has shown that they can be reproduced in modern chemical laboratories with due accuracy (see Principe, *The Secrets of Alchemy*, 137). See also William R. Newman and Lawrence M. Principe, "Alchemy vs. Chemistry: The Etymological Origins of a Historiographic Mistake," *Early Science and Medicine* 3, no. 1 (1998): 32–65.

33. On the revival in the 1770s and 1780s, see Principe, *The Secrets of Alchemy*, 90–91.

34. Principe, *The Secrets of Alchemy*, 91.

35. Goethe, *Dichtung und Wahrheit*, in Goethe, *Werke*, vol. 5 (Insel Verlag, 1998), 305–310. On alchemy and young Goethe, see also Gray, *Goethe the Alchemist*, 54–58; and Karin Figala, "Goethe," in *Alchemie*, ed. Claus Priesner and Karin Figala (C. H. Beck, 1998), 154–57.

36. Goethe, *Dichtung und Wahrheit*, 308: "Wodurch die Natur, wenn auch vielleicht auf phantastische Weise, in einer schönen Verknüpfung dargestellt wird."

37. Goethe, *Dichtung und Wahrheit*, 309–310.

38. For an overview of Goethe's investment in the natural sciences, see Manfred Wenzel, ed., *Goethe-Handbuch: Supplemente. Band 2: Naturwissenschaften* (J. B. Metzler, 2012); and Jonas Maatsch, ed., *Morphologie und Moderne* (De Gruyter, 2014). For the notion of *Naturphilosophie*, see Robert J. Richards, *The Romantic Conception of Life: Science and Philosophy in the Age of Goethe* (University of Chicago Press, 2004), 9–11; on Goethe and the philosophical tenets of *Naturphilosophie*, see Dalia Nassar, "From a Philosophy of Self to a Philosophy of Nature: Goethe and the Development of Schelling's *Naturphilosophie*," *Archiv für Geschichte der Philosophie* 92 (2010): 304–21; and Eckart Föster, "Goethe und die Idee einer Naturphilosophie," in Maatsch, *Morphologie und Moderne*, 43–56, where Goethe's understanding of the principles of nature is discussed in light of Kant's philosophy.

39. See Föster, "Goethe und die Idee einer Naturphilosophie," 47–48. What Goethe achieves through his botanical project is, from a philosophical perspective, "eine Methodologie des intuitiven Verstandes" (53). See also Jonas Maatsch, "Morphologie und Moderne. Zur Einleitung," in Maatsch, *Morphologie und Moderne*, 1–16.

40. This "discovery" (which was actually a rediscovery, since Vesalius had already known of it about two centuries earlier) had the function of reintegrating the allegedly exceptional anatomy of humans, supposedly devoid of this bone, into that of the whole vertebrate animal world. While the novelty of such a discovery was a misrepresentation on the part of Goethe, it was significant in the intellectual aspirations that it drew on; see Ryan Feigenbaum, "Toward a Nonanthropocentric Vision of Nature: Goethe's Discovery of the Intermaxillary Bone," *Goethe Yearbook* 22 (2015): 73–93.

41. On morphology as intuition and "anschauliches Denken," see Maatsch, "Morphologie und Moderne," 2, 7.

42. Lémery observed several kinds of crystallization that were plant-shaped. A distinct, crucial topic in alchemical tradition, the so-called philosophers' tree, was supposedly part of the process to obtain the philosophers' stone for those alchemists defined as "mercurialists." Lawrence Principe has demonstrated the process that leads to the actual chemical reactions leading to a perfectly tree-shaped substance following the instructions provided by Starkey; see Principe, *The Secrets of Alchemy*, 162–66.

43. Goethe wrote to Charlotte von Stein about this book in the summer of 1786. The letter is in Johann Wolfgang Goethe, *Briefe. Historisch-kritische Ausgabe*, ed. Volker Giel (Akademie Verlag, 2010), 207. See also Figala, "Goethe," 157.

44. Following the reassessment offered by Newman and Principe of the history of chymistry up to the eighteenth century (Newman and Principe, "Alchemy vs. Chemistry"), it may be best to look at Goethe's interests as spread along a continuum along which chemical and alchemical traditions were not as neatly separated as we would have it today. In fact, this discontinuity is a later historiographic product (and often the result of a historiographical bias). On the shift at the end of the seventeenth century, when "alchemy" starts to specialize as a definition for metal transmutation, see "Alchemy vs. Chemistry," 39. Dorothea Kuhn fittingly speaks of Goethe's alchemic interests in an essay dedicated to chemistry: Dorothea Kuhn, "Goethe und die Chemie," *Medizinhistorisches Journal* 7, no. 4 (1972): 264–78, esp. 266.

45. Figala, "Goethe," 155.

46. For a concrete explanation of such allegorical images, see Principe, *The Secrets of Alchemy*, 73–81; on the meaning of the hermaphrodite symbology, 78; on the *Rebis*, 122.

47. In recounting years later these events, he defines these early alchemic interests as eccentricities, "Seltsamkeiten" (Goethe, *Dichtung und Wahrheit*, 308). In the same text, while evoking his reading of another alchemic text (Georg von Welling's

Opus Mago-cabbalisticum, published in 1735), Goethe complains about its obscurity (343). Most research on Goethe's scientific interests usually draws a line between Goethe's early interest in and later distance from alchemy understood as hermetic tradition and the rest of his scientific endeavors, including his interests in chemistry. See, for example, Wenzel, "Alchimie," in *Goethe-Handbuch*, 297; and Christa Habrich, "Von der Alchemie zur Förderung der chemischen Wissenschaft und Technik: Goethe zwischen hermetischem Denken und Pragmatismus," in *Von der Pansophie zur Weltweisheit: Goethes analogisch-philosophische Konzepte*, ed. Hans-Jürgen Schrader, Katharine Weder, and Johannes Anderegg (Max Niemeyer Verlag, 2004; De Gruyter, 2011, reprint of 2004 edition), 9–30, esp. 11, 14. Habrich emphasizes the two fields as competing and invested in a friction in Goethe's double experience as state official and poet. Even so, Habrich acknowledges the persistent presence (*basso continuo*) of hermetic, alchemic elements in Goethe's work until the end (26).

48. The book was first published in 1572. Theophrastus Paracelsus, *De natura rerum*, in *Sämtliche Werke in 14 Bänden*, vol. 11, ed. Karl Sudhoff (Oldenbourg, 1928).

49. Paracelsus, *De natura rerum*, 316–18, 317. English translation of the passage in William R. Newman, "The Homunculus and His Forebears: Wonders of Art and Nature," in *Natural Particulars: Nature and the Disciplines in Renaissance Europe*, ed. Anthony Grafton and Nancy G. Siraisi (MIT Press, 1999), 321–45, 328–29. In the quoted passage, I have added in brackets some relevant German expressions from the German edition. On the philological debate around the authenticity of *De natura rerum*, see William R. Newman, *Promethean Ambitions* (University of Chicago Press, 2004), 199.

50. "The possibility of producing life in the laboratory did not appear problematic for medieval and early modern thinkers. The spontaneous origin of life from nonliving matter was considered a matter of course—a rotting bull carcass produced bees, and putrefying mud generated worms and insects." Principe, *The Secrets of Alchemy*, 132.

51. See Newman, "Homunculus and His Forebears," 327.

52. Annie Amartin-Serin, *La Création Défiée: l'Homme Fabriqué dans la Littérature* (Presses universitaires de France, 1996), 32.

53. I am modernizing the original spelling of words in Sudhoff's edition of Paracelsus for clarity. Paracelsus, *De natura rerum*, 317.

54. Newman, "Homunculus and His Forebears," 330–33. On Paracelsus's sources, see also Nabil Mesh-Hadi, *Die Einschätzung der Alchemie in Faust-Deutungen* (Peter Lang, 1979), 185.

55. Paracelsus, *De natura rerum*, 317.

56. Paracelsus, *De natura rerum*, 318.

57. Paracelsus, *De natura rerum*, 319.

58. See Müller, *Die Figur des Homunculus*, 16.

59. Newman, "Homunculus and His Forebears," 321.

60. Principe, *The Secrets of Alchemy*, 74–75. As Principe explains (and demystifies): "The idea or sight of two substances reacting and combining to form a third easily suggests the image of a marital couple to an imaginative mind practiced in drawing metaphors" (77).

61. Though not named as such in the text, philological research has identified annotated manuscripts that bear this definition in the margins (see Newman, "Homunculus and His Forebears" 321; and Gray, *Goethe the Alchemist*, 208).

62. Newman, "Homunculus and His Forebears" 322: "an allegory of spiritual regeneration with the aim of charming the reader rather than teaching him to be Frankenstein."

63. Newman, "Homunculus and His Forebears," 322.

64. Johann Valentin Andreae, *Die Chymische Hochzeit des Christian Rosenkreuz: Anno 1459*, ed. Walter Weber and Rudolf Steiner (Zbinden, 1978), 104 (my translation here and after).

65. Andreae, *Die Chymische Hochzeit*, 105.

66. Andreae, *Die Chymische Hochzeit*, 105.

67. See, for example, Rudolf Steiner's interpretation of Christian Rosenkreuz as "Geistsucher," as well as the interpretation of the sixth day, when homunculi are created, as the moment of initiation, when the protagonist finds out the way to convert the dead powers of knowledge into supernatural intuition (in Andreae, *Die Chymische Hochzeit*, 156).

68. Newman suggests the latter case. Other interpreters insist on a fundamental ambivalence as to the outcome of alchemical experimentation toward the philosophers' stone or the homunculus. For example, for Gray, in *Goethe the Alchemist*, "it is quite impossible to determine here whether the Stone or the homunculus is intended. . . . There seems to be no question of keeping the two symbols apart; they slide into one another with every reference" (210). Gray is radical in concluding that the homunculus was ultimately a synonym for the stone, even though clearly some alchemists believed in the actual possibility of generating life, starting with Paracelsus (211).

69. Beretta, *The Alchemy of Glass*, 84. As for alchemic texts articulating this distinction, the aforementioned *Rosarium Philosophorum*, for example, explicitly conceives of alchemy as dealing only with nature: "As much as its names are different, yet the substance is always only one, and nothing can be introduced in nature that does not already belong to it." *Rosarium Philosophorum, Band 1: Faksimile der illustrierten Erstausgabe, Frankfurt, 1550*, ed. Joachim Telle, Lutz Claren, and Joachim Huber, trans. Lutz Claren, vol. 1 (VCH Verlagsgesellschaft, 1992), 5 (my translation).

70. For Müller, the dramatic function of Homunculus consists precisely in the tension between bodilessness and embodiment. See Müller, *Die Figur des Homunculus*, 17.

71. Habicht's edition was based on Galland's version. See Katharina Mommsen, *Goethe und 1001 Nacht* (Suhrkamp, 1960), 155–56.

72. For a study of Goethe's sources in fables, see Mommsen, *Goethe und 1001 Nacht*; on Helena, see Mommsen, *Natur- und Fabelreich*, 10–13.

73. Goethe to art historian Sulpiz Boisserée, Sept. 8, 1831, quoted in Arendt, "Homunculus," 267.

74. Other possible sources are Friedrich de la Motte Fouqué, *Das Galgenmännlein*; Carl Wilhelm Salice Comtessa, *Magister Röslein*; and Stevenson, *The Bottle Imp* (Arendt, "Homunculus," 267). One interesting case would be the novel *Le Diable boiteux* (1707), by Alain-René Lesage, which tells the story of the devil Asmodeus's liberation from a glass bottle at the hands of the student Don Cleophas. In gratitude, the devil shows the student the private domestic lives of the residents of Madrid, as if there were no walls. In other words, the spirit in the bottle once again possesses the gift of transparent vision (and is able to bestow it on others). See Manfred Schneider, *Transparenztraum: Literatur, Medien und das Unmögliche* (Matthes & Seitz, 2013), 95.

75. E. T. A. Hoffmann, *Tales of E. T. A. Hoffmann*, ed. and trans. Leonard J. Kent and Elizabeth C. Knight (University of Chicago Press, 1969), 76–77. The original version is in E. T. A. Hoffmann, *Der Goldene Topf*, in *Fantasietücke*, ed. Hartmut Steinecke (Deutscher Klassiker Verlag, 2006), 303.

76. The passage from the journals is in Johann Wolfgang von Goethe, *Tagebücher*, in *Sämtliche Werke in 40 Bänden*, vol. 37 (II. Abteilung), ed. Horst Fleig (Deutscher Klassiker Verlag, 1993), 478 (my translation). The mistake in the title is Goethe's ("Den goldnen Becher [*sic*!] angefangen zu lesen").

77. Wilhelm Hauff, *Das kalte Herz* (Insel Verlag, 1989).

78. "A little old man in a black doublet and red stockings, with a great hat upon his head. He had a delicate, kindly little face and a beard as fine as spun cobwebs. He was smoking—strange to behold—from a pipe of blue glass; and as Peter stepped closer, he saw to his astonishment that the little man's clothes, shoes, and hat were also made of colored glass. Yet it was supple, as if still hot, for it clung like cloth to every movement of the little man." Hauff, *Das kalte Herz*, 32 (my translation).

79. "On several wooden ledges stood glasses filled with a clear liquid, and in each of these glasses lay a heart. Labels were also attached to the glasses, with names written on them, which Peter read with curiosity." Hauff, *Das kalte Herz*, 50 (my translation).

80. Hauff, *Das kalte Herz*, 34 (my translation).

81. *Faust I & II*, 199; *Faust*, 238 (my emphasis).

82. On the multilayered satirical components of *Faust II*, see Wilhelm Voßkamp, "Was niemand hören wollte: Aspekte satirischer Kritik der Moderne im *Faust II*," in *Satiren der europäischen Aufklärung*, ed. Gideon Stiening and Friedrich Vollhardt (De Gruyter, 2026), 165–74, esp. 166 and 173 on the caricatural critique of modernity.

83. *Faust I & II*, 199; *Faust*, 238.

84. *Faust I & II*, 200; *Faust*, 238. The exchange that follows between Thales and Anaxagoras (7851–7950) stages the two pre-Socratic philosophies of nature: Thales's

theory of the origin of life from water and Anaxagoras's theory of the origin of life from volcanic activity. These two pre-Socratics closely echo a debate that was current at the beginning of the nineteenth century between so-called Vulcanists and Neptunists, on the origins of Earth (or orogenesis). On the presence of geology in this scene and comprehensively in Goethe's work, see Engelhardt, *Goethe im Gespräch*, esp. 361.

85. *Faust I & II*, 200; *Faust*, 238.

86. *Faust I & II*, 204; *Faust*, 244.

87. *Faust I & II*, 206; *Faust*, 246.

88. *Faust I & II*, 207; *Faust*, 247.

89. *Faust I & II*, 209; *Faust*, 250.

90. *Faust I & II*, 209–10; *Faust*, 250.

91. Kerényi, *Das Ägäische Fest*, 189: "Homunculus' Abenteuer ist das Mysterium des Entstehens."

92. Kerényi, *Das Ägäische Fest*, 187.

93. Jung's reading, some years after Kerényi's, offers a different "mythological" genealogy of *Faust* and alchemy: in both narratives, Jung retrieves a topography of the collective unconscious, imbued with a therapeutic function (Jung, *Psychologie und Alchemie*, 29).

94. Kerényi, *Das Ägäische Fest*, 187.

95. Principe, *The Secrets of Alchemy*, 78–79.

96. Kerényi, *Das Ägäische Fest*, 188: "Sich selbst erzeugt."

97. *Faust I & II*, 210; *Faust*, 250.

98. *Faust I & II*, 211; *Faust*, 252.

99. *Faust I & II*, 211; *Faust*, 252.

100. *Faust I & II*, 214; *Faust*, 255.

101. *Faust I & II*, 214; *Faust*, 255.

102. *Faust I & II*, 215; *Faust*, 256.

103. *Faust I & II*, 215; *Faust*, 256.

104. *Faust I & II*, 215; *Faust*, 256.

105. Hilgers, *Entelechie, Monade und Metamorphose*, 225.

106. On the notion of becoming as the central theme of the drama, see Dorothea Hölscher-Lohmeyer, "Natur und Gedächtnis: Reflexionen über die Klassische Walpurgisnacht," in *Aufsätze zu Goethes "Faust II,"* ed. Werner Keller (Wissenschaftliche Buchgesellschaft, 1992), 94–122, esp. 110.

107. Hölscher-Lohmeyer, "Natur und Gedächtnis," 115. Müller defines *Entstehung* as "Verwandlung": "Entstehen wird nahezu identisch mit verwandeln" (Müller, *Die Figur des Homunculus*, 25).

108. Hölscher-Lohmeyer, "Natur und Gedächtnis," 115 (my translation).

109. See David Wellbery, introduction to *Faust I & II*, xi–xxiv, xvii. Wellbery argues for a parallelism also based on the symmetry between two scenes: Faust's near-suicide and Homunculus's self-sacrifice, both of which involve the question of embodiment. Whereas "Faust's suicide fantasy imagines release from the clunky

encumbrance of corporeality . . . Homunculus' self-sacrifice moves in the opposite direction"; that is, "Faust would negate the body through suicidal fiat; Homunculus merges with the source of living corporeal forms" (xxii).

110. Müller, *Die Figur des Homunculus*, 28–29.

111. Müller, *Die Figur des Homunculus*, 20.

112. Mommsen, *Natur- und Fabelreich*, 184. Arguing for this contrast, Mommsen also points toward the implausibility of Helena's resurrection by emphasizing the degree of unreality that Goethe achieves in the "phantasmagoric" atmosphere of her appearances on the stage (91).

113. Mommsen, *Natur- und Fabelreich*, 236.

114. Mommsen lists what she sees as the main questions in *Faust II*: "What can humans expect from nature and art if they are oriented toward the great epochs of the past and strive for possibilities to improve themselves? How can a model extracted from history become productive in the present? Does the past ever repeat itself?" (Mommsen, *Natur- und Fabelreich*, 243 [my translation]).

115. Quoted in Mommsen, *Natur- und Fabelreich*, 168.

116. For a detailed reconstruction of the genealogy of the notion in Goethe's work, see Hilgers, *Entelechie, Monade und Metamorphose*.

117. Hilgers, *Entelechie, Monade und Metamorphose*, 55, 58n38.

118. On the unity governing his epistemology beyond his scientific interests, see Goethe's own words in his *Hefte zur Naturwissenschaft* (1817): "Nowhere was one willing to admit that science and poetry might be compatible. People forgot that science had developed out of poetry; they did not consider that, after a turn of the ages, the two might once again meet amicably—to their mutual advantage—at a higher level." Johann Wolfgang von Goethe, "Schicksal der Druckschrift," in Goethe, *Werke. Hamburger Ausgabe*, vol. 13, ed. Dorothea Kuhn and Rike Wankenmüller (C. H. Beck, 1981), 105 (my translation).

119. Johann Wolfgang von Goethe, *Selected Poems*, trans. Christopher Middleton (Princeton University Press, 1994), 231; Johann Wolfgang von Goethe, *Gedichte und Epen I*, in Goethe, *Werke. Hamburger Ausgabe*, vol. 1, ed. Erich Trunz (C. H. Beck, 1981), 359.

120. Hilgers has interpreted the stanza in close connection to the notion of *Entelechie*, associating it closely with the concept of *daimon* (Hilgers, *Entelechie, Monade und Metamorphose*, 199).

121. Hilgers, *Entelechie, Monade und Metamorphose*, 197.

122. The conversation, from 1828, is in Eckermann, *Conversations*, 46–47; Eckermann, *Gespräche*, 609–10.

123. A further passage in Eckermann's *Gespräche* from 1830 clarifies the relationship between *Entelechie* and *Monade*: "We continued talking on various subjects, till at last we came to the entelecheia. 'The obstinacy of the individual, and the fact that man shakes off what does not suit him,' said Goethe, 'is a proof to me that something of the kind exists.' I had for some minutes thought the same thing, and was about to express it, and hence I was doubly pleased to hear it uttered by Goethe.

'Leibnitz,' he continued, 'had similar thoughts about independent beings, and indeed what we term an entelecheia, he called a monad.'" Eckermann, *Conversations*, 242; Eckermann, *Gespräche*, 361–62.

124. That is, several historians of alchemy (e.g., Gray) see in it just another allegory of the philosophers' stone.

125. *Faust I & II*, 176. "Natürlichem genügt das Weltall kaum / Was künstlich ist, verlangt geschloßnen Raum" (*Faust II*, 211).

126. The expressions, already quoted above, are from the conversation reported in Eckermann, *Gespräche*, 339.

127. For the suggestion of Goethe's Homunculus as a figure of the posthuman, see also Kevin Liggieri, "Am Ende hängen wir doch ab / Von Kreaturen, die wir machten: Goethes Homunculus als Modell der Entschleunigung," in *Technische Beschleunigung—Ästhetische Verlangsamung? Mobile Inszenierung in Literatur, Film, Musik, Alltag und Politik*, ed. Jan Röhnert (Böhlau, 2015), 260–73.

Chapter 4

1. An accurate account, originally provided by Bruno Gebhard, a former employee of the museum, is in Martin Roth, "Menschenökonomie oder der Mensch als technisches und künstlerisches Meisterwerk," in *Der Gläserne Mensch—eine Sensation: Zur Kulturgeschichte eines Ausstellungsobjekts*, ed. Rosmarie Beier-de Haan and Martin Roth (G. Hatje, 1990), 39–67, 41. See also Klaus Vogel, "The Transparent Man—Some Comments on the History of a Symbol," in *Manifesting Medicine: Bodies and Machines*, ed. Robert Bud, Bernard Finn, and Helmuth Trischler (Harwood Academic, 1999), 45. The most recent and updated research on the history of these museum objects, with a reconstruction of their complicated whereabouts, is Julia Bienholz-Radtke and Susanne Roeßiger, eds., *Transparent Figures: Researching and Preserving Objects of Plastic* (Deutsches Hygiene-Museum, 2021).

2. Over the next few years, an audio track was added. A voice-over accompanied the visual experience with the descriptions of the main plastic organs at the 1935 exhibition, attended by National Socialist officials, *Das Wunder des Lebens*. A recording of the voice-over was made available to me by the archive staff of the Deutsches Historisches Museum in Berlin. On the audio history of the transparent human, see Manuel Vojtech, "Introduction of the Audio Lecture," in Bienholz-Radtke and Roeßiger, *Transparent Figures*, 232.

3. The factory's name was "Mandelpräparate und Marmeladenfabrik Siemank & Ringelhahn" (Julia Bienholz-Radtke and Maria Lörzel, "History of the Production of the Transparent Figures, 1945–2000," in Bienholz-Radtke and Roeßiger, *Transparent Figures*, 19–33, 20. Tschackert was dismissed by the museum in 1946 because of his membership in the *Nationalsozialistische Deutsche Arbeiterpartei* (NSDAP) (21).

4. Julia Bienholz-Radtke, "National Socialist Propaganda with the Transparent Man," in Bienholz-Radtke and Roeßiger, *Transparent Figures*, 204.

5. On the exact exploits of this object, see Bienholz-Radtke and Manuel Vojtech, "Transparent woman from 1935/36" in Bienholz-Radtke and Roeßiger, *Transparent Figures*, 145–49.

6. Bienholz-Radtke and Lörzel, "History of the Production," 21.

7. Bienholz-Radtke and Lörzel, "History of the Production," 23–24.

8. In the Hygiene Museum of Dresden, the historical artifacts are today exposed in their original, materially aged, form, and newer models are available as didactic tools.

9. I consulted these documents at the Archives of the Hygiene Museum in Dresden. I am thankful to Marion Schneider for assistance with the archival material.

10. As of this writing, the Hygiene Museum in Dresden still hosts the old prototype within a glass case.

11. On the history of the eugenics movement, see Sheila Faith Weiss, "The Race Hygiene Movement in Germany," *Osiris* 3 (1987): 193–236; and Sheila Faith Weiss, "German Eugenics, 1890–1933," in *Deadly Medicine: Creating the Master Race*, eds. Susan D. Bachrach and Dieter Kuntz (United States Holocaust Memorial Museum, 2004), 15–39.

12. Giorgio Agamben, *Homo Sacer: Sovereign Power and Bare Life* (Stanford University Press, 1998), 145–46.

13. On this broader rhetoric of medical exhibitions in the twentieth century, see Ludmilla Jordanova, "'Erkenne Dich selbst!' Reflexionen über medizinische Präsentationen in öffentlichen Ausstellungen," in *Erkenne Dich selbst! Strategien der Sichtbarmachung des Körpers im 20. Jahrhundert*, ed. Sybylla Nikolow (Böhlau Verlag, 2015), 118–30, 121.

14. Bernd Wedemeyer-Kolwe, *Der neue Mensch: Körperkultur im Kaiserreich und in der Weimarer Republik* (Königshausen & Neumann, 2004).

15. The portrayed glass woman, now at the Deutsches Hygiene Museum in Dresden (inv. no. 2002/1 L) was crafted in 1935 on commission from an American factory and then toured the US before returning to Germany in 1988. On the history of this specific model, see Julia Bienholz-Radtke and Manuel Vojtech, "Transparent Woman from 1935/36." The TAM model was developed in 1968 by designer Richard Rush, on commission, apparently without reference to the older German model (Adam Langer, "Science and Sculpture," *Chicago Reader*, Oct. 19, 1989). A feature of the model, which famously makes its appearance on Nirvana's album *In Utero* (1994), is a thinner waist and a tilted posture (Julia Bienholz-Radtke and Anna Zabel, "Competitors Products," in Bienholz-Radtke and Roeßiger, *Transparent Figures*, 253).

16. About this and other functions of the models, see Anna Maerker, "'Wunderbare Vorrichtungen' oder 'nutzloses Spielzeug'? Debatten über den öffentlichen Nutzen der Visualisierung des Körperinneren vom 17. bis zum 19. Jahrhundert," in Nikolow, *Erkenne Dich selbst!*, 132–43.

17. On the *Fabrica* and reconfiguration of the relations among anatomy, medical practice, and humanist book culture, see Sachiko Kusukawa, *Andreas Vesalius: Anatomy and the World of Books* (Reaktion, 2024).

18. See also Uli Linke, *German Bodies. Race and Representation After Hitler* (Routledge, 1999), 9–10. Linke defines these bodies as "dramatic visions of the normal," emphasizing the heroic, virile aesthetic of the arrangement of the corpses.

19. Ara Osterweil and David Baumflek, "Emergent Bodies: Human, All too Human, Posthuman," in *The Anatomy of "Body Worlds": Critical Essays on the Plastinated Cadavers of Gunther von Hagens*, ed. Christine Jespersen, Alicita Rodriguez, and Joseph Starr (McFarland, 2009), 244–45.

20. On "anatomy art," see Bernadette Wegenstein, *Getting Under the Skin: Body and Media Theory* (MIT Press, 2006), 84.

21. This is the argument advanced in Rosemarie Beier, "Der Blick in den Körper: Zur Geschichte des gläsernen Menschen in der Neuzeit," in Beier-de Haan and Roth, *Der Gläserne Mensch*, 13–38, 25–27. Beier's point raises interesting observations on the tradition of anatomic representation in connection with the transparent model of Dresden, but the comparison of the two objects, medially asymmetrical, is problematic to sustain.

22. A similar claim is made by Elena Canadelli, who speaks of a "sensuality of anatomical models" replaced by a different, "translucent," aesthetics in the glass models, which also reintegrates the aesthetics of a human machine: "a machine which needs to be well oiled and maintained." Elena Canadelli, " 'Scientific Peep Show': The Human Body in Contemporary Science Museums," *Nuncius* 26 (2011): 159–84, 166.

23. The claim about "immunity" as central to the experience of *Body Worlds* is raised in Osterweil and Baumflek, "Emergent Bodies," 254.

24. Roth, "Menschenökonomie," 44–45. On Pettenkofer, see Wolfgang G. Locher, *Max von Pettenkofer: Pionier der wissenschaftlichen Hygiene* (Pustet, 2018).

25. Roth, "Menschenökonomie," 46–47.

26. Auguste Forel, *Hygiene of Nerves and Mind in Health and Disease*, trans. Herbert A. Aikins (G. P. Putnam's Sons, 1907), 235. First German edition is from 1903.

27. Forel, *Hygiene of Nerves and Mind*, 236,

28. Forel, *Hygiene of Nerves and Mind*, 236–37 (my emphasis).

29. Roth, "Menschenökonomie." Roth estimates that between 1919 and 1937 1,241 mobile exhibits in Germany and abroad attracted 29.5 million visitors.

30. In German: "Gesundheitspflege, Soziale Fürsorge und Leibesübungen."

31. The German name of the exhibition was "Der Mensch in gesunden und kranken Tagen." Deutsches Hygiene-Museum Archive, Dresden (DHMD 2018/749).

32. " 'Das bist du' wird ihm hier gesagt—und 'Erkenn dich Selbst,' wie es über dem alten Griechentempel in Delphi stand, und wenn auch nur ein Teil der Besucher dadurch aufgewühlt und sich bewusst wird, dass er ohne verständnisvolles Sichversenken in die leiblichen Gegebenheiten weder sich selbst recht verstehen und sein Leben führen noch auch dem Mitmenschen gerecht werden kann, dann ist die schönste Aufgabe dieses 'Menschen' erfüllt." *Die Gesundheitspflege : Ausstellung Stettin 1928* (Bauchwitz 1928). Document from the Deutsches Hygiene-Museum Archive, Dresden.

33. "Know Thyself" was a commonplace motto of anatomical museums that opened in the nineteenth century, across Europe and the United States; see Canadelli, "'Scientific Peep Show,'" 164; and Jordanova, "'Erkenne Dich selbst!'"

34. I was able to gather archival material from the 1931 exhibition but not from the 1930 one.

35. "Die Abteilung 'der Mensch' will Verständnis für Bau und *Tätigkeit* des menschlichen Körpers vermitteln und dadurch *Liebe* zu diesem wundervollen *Kunstwerk* wecken. Die Kenntnis von seiner *staunenswerten Organisation* soll *ein freudiges Körpergefühl* anbahnen als Grundlage einer verständigen und nachhaltigen Gesundheitspflege." *Internationale Hygiene-Ausstellung Dresden 1931*. Document from the Deutsches Hygiene-Museum Archive, Dresden.

36. "Entsprechend der dynamischen Betrachtungsweise unserer Zeit, ist der tätige, arbeitende, lebende Organismus in den Vordergrund gestellt." *Internationale Hygiene-Ausstellung Dresden 1931*. Document from the Deutsches Hygiene-Museum Archive, Dresden.

37. Hans Driesch, *Philosophie des Organischen* (Quelle & Meyer, 1909), 143 ("das Leben ist eine Sache für sich"). On the insufficiency of the mechanistic model for biological phenomena: "There can be no kind of machine within the cell from which the individual arises, because this cell, both in its protoplasm and in its nucleus, has emerged from a long series of divisions that all produced identical products, and because a machine does not remain the same when it is divided" (Es kann keine Art von Maschine in der Zelle geben, aus der das Individuum entsteht, weil diese Zelle sowohl in ihrem Protoplasma wie in ihrem Kern aus einer langen Reihe von Teilungen hervorgegangen ist, die alle gleiche Produkte lieferten, und weil eine Maschine nicht dieselbe bleibt, wenn man sie teilt) (244; my translation).

38. On entelechy, see Driesch, *Philosophie des Organischen*, 10–11. A historical rehabilitation of Driesch's Vitalism against the accusations of "metaphysical heresy" is in Bohang Chen, "A Historico-Logical Re-assessment of Hans Driesch's Vitalism," in *Vitalism and Its Legacy in Twentieth-Century Life Sciences and Philosophy*, ed. Charles T. Wolfe and Christopher Donohue (Springer, 2023), 48–65. Driesch was a pacifist and was later expelled from his academic post by the National Socialists; thereafter, he was committed to research in parapsychology.

39. Ernst Haeckel, *Die Lebenswunder. Gemeinverständliche Studien über biologische Philosophie* (Georg Reimer, 1904). In this text, however, Haeckel briefly dismisses Driesch's Vitalist theory and his alleged arrogance. The connection between Nazism and Haeckel is the subject of the classic study by Daniel Gasman, *The Scientific Origins of National Socialism: Social Darwinism in Ernst Haeckel and the German Monist League* (Macdonald, 1971). On Haeckel and Vitalism, as well as on a reassessment of Haeckel's influence on Nazi ideology, see Cat Moir, "What Is Living and What Is Dead in Political Vitalism?," in Wolfe and Donohue, *Vitalism and Its Legacy*, 239–61, esp. 242–46.

40. See Daniel Kevles, "International Eugenics," in *Deadly Medicine: Creating*

the Master Race, ed. Susan D. Bachrach and Dieter Kuntz (United States Holocaust Memorial Museum, 2004), 41–60, 44. See also Sheila Faith Weiss, "German Eugenics," 17: "Eugenics in Germany originated in the late nineteenth century as a reform movement looking for a scientific solution to social problems created by the nation's rapid industrialization and urbanization." A more detailed account is in Peter Weingart, Jürgen Kroll, and Kurt Bayertz, *Rasse, Blut und Gene: Geschichte der Eugenik und Rassenhygiene in Deutschland* (Suhrkamp, 1988).

41. Kevles, "International Eugenics," 51, 55.

42. The Society was first defined as "International," later as "German" (Weiss, "German Eugenics," 19).

43. As Weiß extensively argues in her study, the link between the German eugenics movement and Nazi eugenics policies is historically more complicated than that of mere derivation: "It would be misleading to view racial hygiene in Germany at this time as a mere prelude to 'Nazi eugenics.' The early German racial hygiene movement was preoccupied with eugenics issues common in most other countries with similar movements" (Weiss, "German Eugenics," 19).

Ploetz promoted, for example, Aryan-Jew marriages as advantageous to the biological and social improvement of the race, and in the early years of 1900, members of the *Gesellschaft für Rassenhygiene* (Society for race hygiene) were also Jews. The "master-race," for Ploetz, was loosely defined as the "white" one, and more specifically, it was intended in its socioeconomical "fittest" variant. Ploetz did, in any case, embrace the "Aryan supremacy mystique," unlike Schallmayer (Weiss, "The Race Hygiene Movement," 202–9, 218). See also Kevles, "International Eugenics," 41: "Eugenics is rightly associated with the brutalities of the Nazis, but it was rooted in the pervasive social Darwinism of the late nineteenth century."

44. For Ploetz, race is a "designation of a group of human beings living since generations with respect to their bodily and mental qualities." Alfred Ploetz, *Die Tüchtigkeit unsrer Rasse und der Schutz der Schwachen: Ein Versuch über Rassenhygiene und ihr Verhältnis zu den humanen Idealen, besonders zum Sozialismus* (Fischer, 1895), 2 (my translation). Regarding the extension of the term *hygiene* to the concept of race, Ploetz writes: "Social hygiene and race hygiene are not to be confused. Social hygiene has as a direct aim the wellbeing of the single individual, while race hygiene the wellbeing of a community in time" (3). And later on, coming full circle: "One could then speak of the hygiene of one nation, of one race in a narrower sense, or of the whole human race" (5).

45. Weiss, "The Race Hygiene Movement," 204–5.

46. Weiss, "The Race Hygiene Movement," 218. At the end of the 1920s in Germany, *Rassenhygiene* was often replaced by *Eugenik*. With the rise of the Nazis to power, the institutions of eugenics lost any independence, and the term *Eugenik* was instrumentally replaced by the earlier *Rassenhygiene*.

47. "Mankind must be divided into about two halves: a superior, more socially useful, sounder, or happier, and an inferior, less socially useful, less sound and happy. . . . He who undoubtedly belongs, together with his ancestors, to the upper

half should multiply vigorously; he who belongs no less undoubtedly to the inferior half should hold back, especially if he is incompetent, unhappy, and socially injurious through mental disturbances, crimes, and nervous diseases . . . and finally he who stands in the middle should take care to be moderate in the multiplication of his kind" (Forel, *Hygiene of Nerves*, 278).

48. Forel, *Hygiene of Nerves*, 278.

49. Robert Proctor, *Racial Hygiene: Medicine Under the Nazis* (Harvard University Press, 1988), 19–20; Weiss, "The Race Hygiene Movement," 212.

50. Berlin was the center of a comparatively more progressive core of eugenics, which found an institutional expression in the 1926 Deutscher Bund für Volksaufartung und Erbkunde (German Association for the improvement of the population and the study of heredity). In 1927, however, the Kaiser Wilhelm Institut für Anthropologie, menschliche Erblehre und Eugenik (Institute for anthropology, human heredity, and eugenics) was founded, directed by the racist anthropologist of Southwest Africa Eugen Fischer.

51. Meanwhile, the genetic research carried on in the institutions dedicated to eugenics was heavily based on the study of twins (Weiss, "The Race Hygiene Movement, 229–32).

52. For example, whereas the practice of euthanasia, documented under the Nazis as a secret operation, was never officially aided or supported by eugenicists, it still found silent complicity among them (Weiss, "The Race Hygiene Movement," 234). According to Weiss, the historical assessment of the eugenicists' direct, full responsibility is not the same as assessing their "omission" or connivance, attested by biological materials from Auschwitz sent to the labs of the Berlin institute. On the highly problematic issue of responsibility in this regard, see the discussion in Weingart, Kroll, and Bayertz, *Rasse, Blut und Gene*, 381.

53. Weiss argues for the first interpretation. It was grafting a "new" ideological content into a preexisting form that Nazis made eugenics their own. Weiss's controversial argument is useful to describe the form of the "insertion" of Nazi ideology or the "incorporation" of preexisting elements within their ideological construct (236).

54. Bernd Sösemann and Marius Lange, *Propaganda: Medien und Öffentlichkeit in der NS-Diktatur* (Franz Steiner Verlag, 2011), 834–36; Bienholz-Radtke, "National Socialist Propaganda," 204.

55. See Hans Mommsen, *From Weimar to Auschwitz* (Princeton University Press, 1991).

56. Max Horkheimer and Theodor W. Adorno, *Dialectic of Enlightenment: Philosophical Fragments*, ed. Gunzelin Schmid Noerr, trans. Edmund Jephcott (Stanford University Press, 2002), 212.

57. For a discursive theory of propaganda, based on the notion of intertextuality, see John Oddo, *Discourse of Propaganda: Case Studies from the Persian Gulf War and the War on Terror* (Pennsylvania State University Press, 2024), esp. 21–25.

58. For an overview of constitutive iconographical motives in the transparent model, see Konrad Wünsche, "Das Bildnis des durchsichtigen Gesunden oder: Die

Wahrhaftigkeit des gläsernen Menschen," in Beier-de Haan and Roth, *Der Gläserne Mensch*, 85–101, 99.

59. Wünsche, "Das Bildnis des durchsichtigen Gesunden," 99.

60. On the Vitalist connections to Nazi ideology, see Georges Canguilhem, *Knowledge of Life*, ed. Paola Marrati and Todd Meyers (Fordham University Press, 2008), 72–73.

61. *Internationale Hygiene-Ausstellung Dresden 1931: amtlicher Führer* (Dresden, 1931). (my emphasis):

> Wie in einem *Märchen* dringt das Auge in *verborgene Tiefen des Körpers* ein und durchschaut in den durchsichtigen Präparaten die räumliche Struktur der Organe so klar, wie sie nicht einmal die Röntgenstrahlen enthüllen können. Hier offenbart sich die Formenschönheit des Körpers und seiner Einzelteile als Sinnbild der *vollendeten Harmonie*, die in den *Gesetzen des Leben* waltet. In sich selbst findet der Mensch *die grössten Wunder der Welt*, wenn er, worauf das Wahrzeichen des Museums hinweist, mit wachem Auge um sich und in sich blickt, und daraus wird ihm Achtung vor seinem Körper erwachsen und ein tiefes *Verantwortungsgefühl* für das höchste Gut, das er zu bewahren hat, die Gesundheit.

62. Sebastian Weinert, *Der Körper im Blick: Gesundheitsausstellungen vom späten Kaiserreich bis zum Nationalsozialismus* (De Gruyter, 2017), 296.

63. "Das Wunder des Lebens / Ausstellung Berlin 1935." Document DHMD 2017/1337, Deutsches Hygiene-Museum Archive, Dresden:

> So soll die grosse Schau in zweifacher Richtung wirken; sie soll dem Volksgenossen in leicht verständlicher und höchst einprägsamer Form Wissen um den höchsten Organismus, den die Natur schuf, ums eigene Ich vermitteln. Sie soll aber zugleich die grossen Gesetze, nach denen die Natur waltet, sichtbar machen, und soll unser *Lebensgefühl* vertiefen, unseren *Lebenswillen* bahnweisend stärken. Wer durch die Hallen schreiten will, wird in Bann geschlagen von der Tatsache, dass mit dem Ernst der Ausstellung "Das Wunder des Lebens" ein starkes *Frohgefühl* sich paart, *das Frohgefühl der Lebensfreude* von Menschen die, einem grossen Volke zugehörig, nicht nur Erben einer starken Vergangenheit, sondern auch Träger einer edlen Zukunft sind.

64. See Oddo, *Discourse of Propaganda*, 21.

65. "Die Abteilung 'der Mensch' will Verständnis für Bau und *Tätigkeit* des menschlichen Körpers vermitteln und dadurch *Liebe* zu diesem wundervollen *Kunstwerk* wecken. Die Kenntnis von seiner *staunenswerten Organisation* soll *ein freudiges Körpergefühl* anbahnen als Grundlage einer verständigen und nachhaltigen Gesundheitspflege." *Internationale Hygiene-Ausstellung Dresden 1931* (my emphasis).

66. On the different female glass models, see Julia Bienholz-Radtke, "Transpar-

ent Women: An Exhibition History," in Bienholz-Radtke and Roeßiger, *Transparent Figures*, 41–56.

67. Bienholz-Radtke, "Transparent Women," 43. About health assumptions and indications for women's health within the Nazi era's hygiene exhibitions, see Weinert, *Der Körper im Blick*, 310–13.

68. Driesch, *Philosophie des Organischen*, 30–33.

69. Max Frisch, *Little Diary of a German Journey*, trans. Linda Frazee Baker, *New England Review* 41, no. 1 (2020): 135–44, 137; Max Frisch, "Kleines Tagebuch einer deutschen Reise," in *Gesammelte Werke in zeitlicher Folge*, vol. 1, *1931–1944*, ed. Hans Mayer, with the collaboration of Walter Schmitz (Suhrkamp, 1976), 84–97, 88.

70. Frisch, *Little Diary*, 138; Frisch, "Kleines Tagebuch," 89.

71. Hal Foster, "Armor Fou," *October* 56 (1991): 64–97, 67. See also Hal Foster, "Exquisite Corpses," *Visual Anthropology Review* 7 (1991): 51–61. Foster's psychoanalytically informed interpretation of the fascist "armored body" sees it as a "prosthesis" implemented as protection for a "disrupted body image," which, in its turn, reflects "a ruined ego construction" (Foster, "Armor Fou," 68). His reading draws explicitly on Lacan's study of compensatory forms of identity of a weak male heterosexual ego failing at embodying virility (69n8). For Foster, this aesthetic deeply shapes political fascism, as aggression seems necessary to the (proto)fascist not only for self-definition but also for self-defense, producing, however, self-loathing effects (84). This is the same thesis embraced by Theweleit in Klaus Theweleit, *Male Fantasies*, trans. Stephen Conway (University of Minnesota Press, 1987), to whom Foster also refers ("Armor Fou," 84).

72. Foster, "Armor Fou," 91.

73. Horkheimer and Adorno, *Dialectic of Enlightenment*, 195. This aesthetic is conceived as the outcome of a preliminary reification of the body: "The body cannot be turned back into the envelope of the soul. It remains a cadaver, no matter how trained and fit it may be. The transformation into dead matter, indicated by the affinity of *corpus* to corpse, was a part of the perennial process which turned nature into stuff, material" (194). On the same line of this later analysis is Max Frisch's bitter diagnosis on the Berlin exhibition of 1935: "This exaggerated sense of self is the extreme swing of an inner pendulum, and we are also aware of German self-doubt, which can go so far as self-destruction and is without equal. In short, we are by no means unacquainted with the German soul. What is shocking, however, is this self-glorification, which exalts its own race by casting all others into the dirt" (Frisch, "Kleines Tagebuch," 91).

74. Wedemeyer-Kolwe, *Der neue Mensch*, 19.

75. On the movement and its cultural success, especially toward literary production, see Thorsten Carstensen and Marcel Schmid, eds., *Die Literatur der Lebensreform: Kulturkritik und Aufbruchstimmung um 1900* (Bielefeld University Press, 2016).

76. Although some of the implications of this aesthetic movement could be

easily put to the service of National Socialist ideology, there is no intrinsic relationship between the two, which is reflected, for example, on the complicated biographical vicissitudes of the artist in question when it came to his relation to the NSDAP. For a biography of Höppener, see Gisela Frecot, Johann Friedrich Geist, and Diethart Kerbs, *Fidus, 1868–1948* (Rogner & Bernhard, 1997).

77. The verso of the watercolor in figure 17 explains: "Mit Ätheraugen (Röntgenschwingungen) gesehen 1914," hinting explicitly at the visionary possibilities disclosed by X-Ray technology. I am thankful to Jack Daulton for this information about the picture from his collection.

78. Canadelli called it "a synesthetic museological performance" (Canadelli, "Scientific Peep Show," 169).

79. Foster, "Armor Fou," 51.

80. Horkheimer and Adorno, *Dialectic of Enlightenment*, 195.

81. Anson Rabinbach and Sander L. Gilman, eds., *The Third Reich Sourcebook* (University of California Press, 2013), 496. The passage continues: "But that would also imply that to be German is thus to be logical and, above all, to be true. It is truly a magnificent law, but it is one that everyone must follow and serve in order for it to take force. And this law also provides us with a standard for measuring the value of art because it is congruent with the natural laws and essence of our *Volk*."

82. Rabinbach and Gilman, *The Third Reich Sourcebook*, 498 (my emphasis). I am providing here the original German text for the purpose of my argument. "Die heutige neue Zeit arbeitet an einem neuen Menschentyp. Ungeheure Anstrengungen werden auf unzähligen Gebieten des Lebens vollbracht, um das Volk zu heben, um unsere Männer, Knaben und Jünglinge, die Mädchen und Frauen gesünder und damit kraftvoller und schöner zu gestalten. Und aus dieser Kraft und aus dieser Schönheit strömen ein neues Lebensgefühl, eine neue Lebensfreude!" Adolf Hitler, *Reden und Proklamationen, 1932–1945; Kommentiert von einem Deutschen Zeitgenossen* (Schmidt, 1962), 706–10.

83. In Hitler's speech, these degenerate artists would misrepresent their subjects: "Wiesen blau, Himmel grün, Wolken schwefelgelb usw. empfinden oder, wie sie vielleicht sagen, erleben."

84. The athletic body promoted by the Nazis is celebrated, for example, in the sculptures of Arno Breker.

85. Preserving these cellulose acetate objects is a challenge for conservators, whose best achievable result is slowing down the process of ageing. See Maria Lörzel, "The Practical Conservation of the Transparent Man from 1935 in the Deutsches Hygiene Museum," in Bienholz-Radtke and Roeßiger, *Transparent Figures*, 113–21.

86. For the continuing history of the models, see Bienholz-Radtke and Lörzel, "History of the Production."

Coda

1. For an overview of the field of transparency studies, see the comprehensive overview offered in the volume edited by Emmanuel Alloa and Dieter Thomä, *Transparency, Society and Subjectivity: Critical Perspectives* (Palgrave Macmillan, 2018), 155–76. For a cultural historical study of transparency also dedicated to some of the figures of this book, but within a general conceptual framework, see Manfred Schneider, *Transparenztraum. Literatur, Politik, Medien und das Unmögliche* (Matthes & Seitz, 2013).

2. For the passages of this genealogy, see the different but congruous accounts provided in Vidler, *The Architectural Uncanny*, and Geroulanos, *Transparency in Postwar France*.

3. Anthony Vidler, *The Architectural Uncanny* (MIT Press, 1992), 218.

4. Erich Auerbach, *Mimesis: The Representation of Reality in Western Literature*, trans. Willard R. Trask, with a new introduction by Edward W. Said (Princeton University Press, 2003), 549.

5. Wallace Stevens, "Asides on the Oboe," in *The Collected Poems of Wallace Stevens* (Alfred A. Knopf, 1971), 250. Perceived as antirealist for a long time, Stevens's poetics is today reframed as realist; see Simon Critchley, *Things Merely Are: Philosophy in the Poetry of Wallace Stevens* (Routledge, 2005), 26.

6. Charles Lock, "Tabish Khair: Man of Glass," in *Tabish Khair: A Critical Companion*, ed. Om Prakash Dwivedi (Roman Books, 2013), 30–38, 32. Lock reads Stevens's figure as a response to an Emersonian transcendentalist subject.

7. Stevens, *The Collected Poems*, 251.

8. Virginia Woolf, *On Being Ill* (Paris Press, 2012), 4.

9. The episode first appeared in *World's Finest*, no. 28 (1947) and is reprinted in *Batman: The World's Finest Comics Archives*, vol. 2 (2004).

10. *The Incredible Hulk*, no. 138 (1971).

11. Anne Carson, "The Glass Essay," in *Glass, Irony and God* (New Directions, 1995), 1–38, 37.

12. Carson scholar Elizabeth Sarah Coles describes this predicament: "Glass is the reluctant matter that, depending on where we're standing, either effaces or exposes its mediation—another way of saying that it keeps the idea of mediation open." Elizabeth Sarah Coles, *Anne Carson: The Glass Essayist* (Oxford University Press, 2023), 14.

13. Carson, "The Glass Essay," 34. In the poem, the soul trapped in glass is actually Emily Brontë's, who functions as the epistemic foil of the lyrical subject in search of her own vision.

14. "Glass can either effect metonymies between its two sides or, by staying the eye and denying that it has another side, yield metaphors of reflection." Charles Lock, "Glass Glimpsed: In, On, Through and Beyond," in *Invisibility Studies: Surveillance, Transparency and the Hidden in Contemporary Culture*, ed. Henriette Steiner and Kristin Veel (Peter Lang 2015), 5–24, 19.

15. See Carson, "The Glass Essay," 38:

I saw it was a human body

trying to stand against winds so terrible that the flesh was blowing off the
bones.

. . . .

The wind

was cleansing the bones.
They stood forth silver and necessary.
It was not my body, not a woman's body, it was the body of us all.

16. On the cultural history of the naked truth topos, see Quintus Immisch di Padua, *Nacktheit: Eine Kulturgeschichte des Körpers* (De Gruyter, 2026), esp. 154, 321, 366.

17. For a reading of transparency as "dream," see Schneider, *Transparenztraum*.

18. Geroulanos, "Transparency, Humanism," 21; another poignant definition of transparency as a function of modernity is formulated as "the promise and illusion of modernity as transparency" (155).

19. On Foucault, see Geroulanos, *Transparency in Postwar France*, 300–301.

20. Geroulanos, *Transparency in Postwar France*, 8–9, 21; and Geroulanos, "Transparency, Humanism," 157–58.

21. Geroulanos, *Transparency in Postwar France*, 10.

22. A case in point is Hans-Georg Gadamer's foregrounding of transparency as condition of understanding (see Geroulanos, *Transparency in Postwar France*, 372).

23. Geroulanos describes it as "the new rhetoric of freedom of information and its advocacy of social and institutional transparency" ("Transparency, Humanism," 377).

24. On the contemporary political discussion, see Emmanuel Alloa, "Transparency: A Magic Concept of Modernity," in Alloa and Thomä, *Transparency, Society and Subjectivity*, 21–56, esp. 24; and, in the same volume, Amitai Etzioni, "The Limits of Transparency," 179–201, esp. 179. Etzioni's essay questions the given assumption of political transparency as a "major good."

25. A case in point in American politics would be the refusal to actually produce governmental disclosure but nonetheless claiming public transparency on nonpolitical platforms provided, for example, by social media; see Geroulanos, "Transparency, Humanism," 157.

26. See Geroulanos, *Transparency in Postwar France*, 188.

27. See Dieter Mersch, "Obfuscated Transparency," in Alloa and Thomä, *Transparency, Society and Subjectivity*, 259–82, 275.

28. Mersch, "Obfuscated Transparency," 278: "neither our intentions nor our actions are transparent, nor are the structures of publicity and the organizations that run them. The only thing that is transparent are the traces that we leave behind. Stored in huge data repositories, they can be retrieved at any time in the form of data profiles."

29. As Geroulanos put it, "the liberal dream of transparency has been inverted into a technologically dominated and financially promoted imaginary of self-exposure." Geroulanos, "Transparency, Humanism," 174.

30. Marcel Rosenbach, Cornelia Schmergal, and Hilmar Schmundt, "Der Gläserne Patient," *Der Spiegel*, Dec. 5, 2015, 10–18.

31. Cathy Waldby summarizes this tendency as the overcoming of "the problem of bodily opacity." See Cathy Waldby, *The Visible Human Project: Informatic Bodies and Posthuman Medicine* (Psychology Press, 2000), 14.

32. Bettyann Holtzmann Kevles, *Naked to the Bone: Medical Imaging in the Twentieth Century* (Basic Books, 1997), 2.

33. Kevles, *Naked to the Bone*, 117.

34. Waldby, *The Visible Human Project*, 14.

35. Harvard Dataverse, *Visible Human Project CT Datasets Male and Female*, version 1.0 (Harvard Dataverse, 2023), dataset.

36. Waldby, *The Visible Human Project*, 5.

37. Contemporary 3D animation technologies that employ these images for more holistic visual effects have multiplied, but they do not partake of the database production.

38. Beatrice Alemagna, *Child of Glass*, trans. Claudia Zoe Bedrick (Enchanted Lion, 2019).

39. David Foster Wallace, *Infinite Jest: A Novel* (Little, Brown, 1996), 989.

Index

Cultural Memory in the Present

Ali Benmakhlouf, *The Humanity of Others*
Arvi Särkelä, *Vicious Circles: Disclosing a History of Critique*
Eyal Peretz, *American Medium: A New Film Philosophy*
Gilah Kletenik, *Sovereignty Disrupted: Spinoza and the Disparity of Reality*
Jensen Suther, *True Materialism: Hegelian Marxism and the Modernist Struggle for Freedom*
Jean-Luc Marion, *Cartesian Questions III: Descartes Beneath the Mask of Cartesianism*, translated by Stephen E. Lewis and Stephanie Rumpza
Walter Benjamin, *On Goethe*, edited by Susan Bernstein, Peter Fenves, and Kevin McLaughlin
Severo Sarduy, *Barroco and Other Writings*, translated by Alex Verdolini
Elliot R. Wolfson, *Nocturnal Seeing: Hopelessness of Hope and Philosophical Gnosis in Susan Taubes, Gillian Rose, and Edith Wyschogrod*
David D. Kim, *Arendt's Solidarity: Anti-Semitism and Racism in the Atlantic World*
Hans Joas, *Why the Church?: Self-Optimization or Community of Faith*
Jean-Luc Marion, *Revelation Comes from Elsewhere*, translated by Stephen E. Lewis and Stephanie Rumpza
Peter Sloterdijk, *Out of the World*
Christopher J. Wild, *Descartes' Meditative Turn: Cartesian Thought as Spiritual Practice*
Eli Friedlander, *Walter Benjamin and the Idea of Natural History*
Helmut Puff, *The Antechamber: Toward a History of Waiting*
Raúl E. Zegarra, *A Revolutionary Faith: Liberation Theology Between Public Religion and Public Reason*
David Simpson, *Engaging Violence: Civility and the Reach of Literature*
Michael Steinberg, *The Afterlife of Moses: Exile, Democracy, Renewal*
Alain Badiou, *Badiou by Badiou*, translated by Bruno Bosteels
Eric Song, *Love against Substitution: Seventeenth-Century English Literature and the Meaning of Marriage*
Niklaus Largier, *Figures of Possibility: Aesthetic Experience, Mysticism, and the Play of the Senses*
Mihaela Mihai, *Political Memory and the Aesthetics of Care: The Art of Complicity and Resistance*
Ethan Kleinberg, *Emmanuel Levinas's Talmudic Turn: Philosophy and Jewish Thought*
Willemien Otten, *Thinking Nature and the Nature of Thinking: From Eriugena to Emerson*
Michael Rothberg, *The Implicated Subject: Beyond Victims and Perpetrators*

Hans Ruin, *Being with the Dead: Burial, Ancestral Politics, and the Roots of Historical Consciousness*
Eric Oberle, *Theodor Adorno and the Century of Negative Identity*
David Marriott, *Whither Fanon? Studies in the Blackness of Being*
Reinhart Koselleck, *Sediments of Time: On Possible Histories*, translated and edited by Sean Franzel and Stefan-Ludwig Hoffmann
Devin Singh, *Divine Currency: The Theological Power of Money in the West*
Stefanos Geroulanos, *Transparency in Postwar France: A Critical History of the Present*
Sari Nusseibeh, *The Story of Reason in Islam*
Olivia C. Harrison, *Transcolonial Maghreb: Imagining Palestine in the Era of Decolonialization*
Barbara Vinken, *Flaubert Postsecular: Modernity Crossed Out*
Aishwary Kumar, *Radical Equality: Ambedkar, Gandhi, and the Problem of Democracy*
Simona Forti, *New Demons: Rethinking Power and Evil Today*, translated by Zakiya Hanafi
Joseph Vogl, *The Specter of Capital*
Hans Joas, *Faith as an Option: Possible Futures for Christianity*
Michael D. Gubser, *The Far Reaches: Phenomenology, Ethics, and Social Renewal in Central Europe*
Françoise Davoine, *Mother Folly: A Tale*, translated by Judith G. Miller with a preface by Mieke Bal
Knox Peden, *Spinoza Contra Phenomenology: French Rationalism from Cavaillès to Deleuze*
Elizabeth A. Pritchard, *Religion in Public: Locke's Political Theology*
Ankhi Mukherjee, *What Is a Classic? Postcolonial Rewriting and Invention of the Canon*
Jean-Pierre Dupuy, *The Mark of the Sacred*, translated by M. B. DeBevoise
Henri Atlan, *Fraud: The World of Ona'ah*, translated by Nils F. Schott
Niklas Luhmann, *Theory of Society, Volume 2*, translated by Rhodes Barrett
Ilit Ferber, *Philosophy and Melancholy: Benjamin's Early Reflections on Theater and Language*
Alexandre Lefebvre, *Human Rights as a Way of Life: On Bergson's Political Philosophy*
Theodore W. Jennings, *Jr., Outlaw Justice: The Messianic Politics of Paul*
Alexander Etkind, *Warped Mourning: Stories of the Undead in the Land of the Unburied*
Denis Guénoun, *About Europe: Philosophical Hypotheses*, translated by Christine Irizarry
Maria Boletsi, *Barbarism and Its Discontents*
Sigrid Weigel, *Walter Benjamin: Images, the Creaturely, and the Holy*, translated by Chadwick Truscott Smith
Roberto Esposito, *Living Thought: The Origins and Actuality of Italian Philosophy*, translated by Zakiya Hanafiy
Henri Atlan, *The Sparks of Randomness, Volume 2: The Atheism of Scripture*

For a complete listing of titles in this series, visit the Stanford University Press website, www.sup.org.

Printed and bound by CPI Group (UK) Ltd, Croydon, CR0 4YY

05/10/2026

14975124-0001